NO-DRAMA DISCIPLINE

DANIEL J. SIEGEL, MD, received his medical degree from Harvard University and completed his postgraduate medical education at UCLA, where he is currently a clinical professor at the David Geffen School of Medicine. An award-winning educator, he has been responsible for the publication of dozens of books as author, co-author, or editor, including *Brainstorm: the power and purpose of the teenage brain* and *Mindsight: change your brain and your life*. He is the executive director of the Mindsight Institute, an educational centre for interpersonal neurobiology that combines various fields of science into one framework for understanding human development and the nature of well-being. Visit www. DrDanSiegel.com.

TINA PAYNE BRYSON, PHD, is the co-author (with Dan Siegel) of the bestselling *The Whole-Brain Child,* which has been translated into eighteen languages. She is a pediatric and adolescent psychotherapist, the director of parenting for the Mindsight Institute, and the child-development specialist at Saint Mark's School in Altadena, California. She lives near Los Angeles with her husband and three children. Visit TinaBryson.com.

BY DANIEL J. SIEGEL, MD, AND TINA PAYNE BRYSON, PhD

The Whole-Brain Child

No-Drama Discipline

BY DANIEL J. SIEGEL, MD

Brainstorm

Mindsight

NO-DRAMA DISCIPLINE

THE WHOLE-BRAIN WAY TO CALM THE CHAOS *AND* NURTURE YOUR CHILD'S DEVELOPING MIND

DANIEL J. SIEGEL, MD

AND

TINA PAYNE BRYSON, PhD

SCRIBE

Melbourne • London

Scribe Publications
18–20 Edward St, Brunswick, Victoria 3056, Australia
2 John St, Clerkenwell, London, WC1N 2ES, United Kingdom

Published by Scribe 2015
Reprinted 2015 (three times), 2016, 2017, 2018

This edition published by arrangement with Bantam Books, an imprint of Random House,
a division of Random House LLC, a Penguin Random House Company, New York.

All identifying details, including names, have been changed except for those pertaining to
the authors' family members. This book is not intended as a substitute for advice from a
trained professional.

The moral right of the authors has been asserted.

Illustrations by Tuesday Mourning

Printed and bound in the UK by CPI Group (UK) Ltd, Croydon CR0 4YY

9781925106152 (Australian edition)
9781922247568 (UK edition)
9781925113327 (e-book)

CiP records for this title are available from the British Library and
the National Library of Australia

scribepublications.com.au
scribepublications.co.uk

To the youth of the world, our vital teachers (DJS)

For my parents: my first teachers and my first loves (TPB)

CONTENTS

A Question

A cereal bowl gets thrown across the kitchen, splattering milk and Cheerios all over the wall.

The dog runs in from the backyard and has inexplicably been painted blue.

One of your kids threatens a younger sibling.

You get a call from the principal's office for the third time this month.

What do you do?

Before you answer, we want to ask you to completely forget about everything you know about discipline. Forget what you think the word means, and forget what you've heard about how parents should respond when kids do something they're not supposed to.

Instead, ask yourself a question: Are you open to at least thinking about a different approach to discipline? One that helps you achieve your immediate goals of getting your kids to do the right thing in the moment, as well as your longer-range goals of helping them become good people who are happy, successful, kind, responsible, and even self-disciplined?

If so, this book is for you.

Relational, Low-Drama Discipline: Encouraging Cooperation While Building a Child's Brain

You are not alone.

If you feel at a loss when it comes to getting your kids to argue less or speak more respectfully . . . if you can't figure out how to keep your toddler from climbing up to the top bunk, or get him to put on clothes before answering the front door . . . if you feel frustrated having to utter the same phrase over and over again ("Hurry! You're going to be late for school!") or to engage in *another* battle over bedtime or homework or screen time . . . if you've experienced any of these frustrations, you are not alone.

In fact, you're not even unusual. You know what you are? A parent. A human being, and a parent.

It's hard to figure out how to discipline our kids. It just is. All too often it goes like this: They do something they shouldn't do. We get mad. They get upset. Tears flow. (Sometimes the tears belong to the kids.)

It's exhausting. It's infuriating. All the drama, the yelling, the hurt feelings, the guilt, the heartache, the disconnection.

Do you ever find yourself asking, after an especially agonizing interaction with your kids, "Can't I do better than this? Can't I handle myself better, and be a more effective parent? Can't I discipline in ways that calm the situation rather than create more chaos?" You want the bad behavior to stop, but you want to respond in a way that values and enhances your relationship with your children. You want to build your relationship, not damage it. You want to create less drama, not more.

You can.

In fact, that's the central message of this book: *You really can discipline in a way that's full of respect and nurturing, but that also maintains clear and consistent boundaries. In other words, you can do better.* You can discipline in a way that's high on relationship, high on respect, and low on drama and conflict—and in the process, you can foster development that builds good relationship skills and improves your children's ability to make good decisions, think about others, and act in ways that prepare them for lifelong success and happiness.

We've talked to thousands and thousands of parents all over the world, teaching them basics about the brain and how it affects their relationship with their kids, and we've seen how hungry parents are to learn to address children's behavior in ways that are more respectful *and* more effective. Parents are tired of yelling so much, tired of seeing their kids get so upset, tired of their children continuing to misbehave. These parents know the kind of discipline they *don't* want to use, but they don't know *what to do* instead. They want to discipline in a kind and loving way, but they feel exhausted and overwhelmed when it comes to actually getting their kids to do what they're supposed to do. They want discipline that works and that they feel good about.

In this book, we'll introduce you to what we call a No-Drama, Whole-Brain approach to discipline, offering principles and strategies that will remove most of the drama and high emotions that so typically characterize discipline. As a result, your life as a parent will be easier and your parenting will become more effective. More important, you'll create connections in your children's brains that build emotional and social skills that will serve them now and throughout their entire life—all while strengthening your relationship with them.

What we hope you'll discover is that the moments when discipline is called for are actually some of the most important moments of parenting, times when we have the opportunity to shape our children most powerfully. When these challenges arise—and they will—you'll be able to look at them not merely as dreaded discipline situations full of anger and frustration and drama, but as opportunities to connect with your children and redirect them toward behavior that better serves them and your whole family.

If you are an educator, therapist, or coach who is also responsible for the growth and well-being of children, you will find that these techniques work just as well for your students, patients and clients, or teams. Recent discoveries about the brain give us deep insights into the children we care for, what they need, and how to discipline them in ways that foster optimal development. We've written this book for anyone who cares for a child and is interested in loving, scientifically informed, effective strategies to help children grow well. We'll use the word "parent" throughout the book, but if you're a grandparent, a teacher, or some other significant person in the life of a child, this book is also for you. Our lives are more meaningful with collaboration, and this joining together can begin with the many adults who cooperate in the nurturing of a child in the earliest days of life onward. We hope all children have many caregivers in their lives who are intentional about how they interact with them and, when necessary, discipline them in ways that build skills and enhance their relationship.

Reclaiming the Word "Discipline"

Let's begin with the actual goal of discipline. When your child misbehaves, what do you want to accomplish? Are consequences your ultimate goal? In other words, is the objective to punish?

Of course not. When we're angry, we may *feel* like we want to punish our child. Irritation, impatience, frustration, or just being unsure can make us feel that. It's totally understandable—even common. But once we've calmed down and cleaned the raw egg out of everyone's hair, we know that giving consequences is not our ultimate goal.

So what *do* we want? What *is* the goal of discipline?

Well, let's start with a formal definition. The word "discipline" comes directly from the Latin word *disciplina,* which was used as far back as the eleventh century to mean teaching, learning, and giving instruction. So, from its inception in the English language, "discipline" has meant "to teach."

These days, most people associate only punishment or consequences with the practice of discipline. It's like the mother with the eighteen-month-old son who asked Dan: "I'm doing a lot of teaching with Sam, but when do I start disciplining him?" The mother saw that she needed to address her son's behaviors, and she assumed that punishment is what discipline is meant to be.

As you read the rest of this book, we want you to keep in mind what Dan explained: that whenever we discipline our kids, our overall goal is not to punish or to give a consequence, but to teach. The root of "discipline" is the word *disciple,* which means "student," "pupil," and "learner." A disciple, the one receiving discipline, is not a prisoner or recipient of punishment, but one who is learning through instruction. Punishment might shut down a behavior in the short term, but teaching offers skills that last a lifetime.

We thought a lot about whether we even wanted to use the word "discipline" in our title. We weren't sure what to call this practice of setting limits while still being emotionally attuned to our children, this approach that centers on teaching and working with our kids to help them build the skills to make good choices. We decided that we want to reclaim the word "discipline," along with its original meaning. We want to completely reframe the whole discussion and differentiate *discipline* from *punishment.*

Essentially, we want caregivers to begin to think of discipline as one of the most loving and nurturing things we can do for kids. Our children need to *learn* skills like inhibiting impulses, managing big angry feelings, and considering the impact of their behavior on others. Learning these essentials of life and relationships is what they need, and if you can provide it for them, you'll be offering a significant gift not only to your children, but to your whole family and even the rest of the world. Seriously. This is not mere hyperbole. No-Drama Discipline, as we'll describe it in the coming pages, will help your kids become the people they are meant to be, improving their ability to control themselves, respect others, participate in deep relationships, and live moral and ethical lives. Just think, then, about the generational impact that will have as they grow up with these gifts and abilities, and raise children of their own, who can then pass on these same gifts to future generations!

It begins with rethinking what discipline really means, reclaiming it as a term that's not about punishment or control, but about teaching and skill building—and doing so from a place of love, respect, and emotional connection.

The Dual Goals of No-Drama Discipline

Effective discipline aims for two primary goals. The first is obviously to get our kids to cooperate and do the right thing. In the heat of the moment, when our child is throwing a toy in a restaurant or being rude or refusing to do homework, we simply want her to act like she's supposed to. We want her to stop throwing the toy. We want her to communicate respectfully. We want her to get her homework done.

With a small child, achieving the first goal, cooperation, might involve getting him to hold your hand as he crosses the street, or helping him put down the bottle of olive oil he's swinging like a baseball bat in aisle 4 at the grocery store. For an older child it might mean problem-solving with him to do his chores in a more timely fashion, or discussing how his sister might feel about the phrase "fat-butted lonely girl."

You'll hear us say it repeatedly throughout the book: every child is different, and no parenting approach or strategy will work every time. But the most obvious goal in all of these situations is to elicit cooperation and to help a child behave in ways that are acceptable (like using kind words, or putting dirty clothes in the hamper) and avoid behaviors that aren't (like hitting, or touching the gum someone left under the table at the library). This is the short-term goal of discipline.

For many people, that's the only goal: gaining immediate cooperation. They want their kids to stop doing something they shouldn't be doing or begin doing something they should be doing. That's why we so often hear parents use phrases like "Stop it *now*!" and the timeless "Because I said so!"

But really, we want more than mere cooperation, don't we? Of course we want to prevent the breakfast spoon from becoming a weapon. Of course we want to promote kind and respectful actions, and reduce the insults and belligerence.

But there's a second goal that's just as important, and whereas getting cooperation is the short-term objective, this second goal is more long-term. It focuses on instructing our children in ways that develop skills and the capacity to resiliently handle challenging situations, frustrations, and emotional storms that might make them lose control. These are the internal skills that can be generalized beyond the immediate behavior in the moment and then used not only now, but later in a variety of situations. This internal, second major goal of discipline is about helping them develop self-control and a moral compass, so that even when authority figures aren't around, they are thoughtful and conscientious. It's about helping them grow up and become kind and responsible people who can enjoy successful relationships and meaningful lives.

We call this a Whole-Brain approach to discipline because, as we'll explain, when we use the whole of our own brain as parents, we can focus on both the immediate external teachings and the long-term internal lessons. And when our children receive this form of intentional teaching, they, too, come to use their whole brains.

Over the generations, countless theories have cropped up about how to help our children "grow up right." There was the "spare the rod and spoil the child" school, and its opposite, the "free to be you and me" school. But in the last twenty years or so, during what's been called "the decade of the brain" and the years that have followed, scientists have discovered an immense amount of information about the way the brain works, and it has plenty to tell us about loving, respectful, consistent, effective discipline.

We now know that the way to help a child develop optimally is to help create connections in her brain—her whole brain—that develop skills that lead to better relationships, better mental health, and more meaningful lives. You could call it brain sculpting, or brain nourishing, or brain building. Whatever phrase you prefer, the point is crucial, and thrilling: as a result of the words we use and the actions we take, children's brains will actually change, and be built, as they undergo new experiences.

Effective discipline means that we're not only stopping a bad behavior or promoting a good one, but also teaching skills and nurturing the connections in our children's brains that will help them make better decisions and handle themselves well in the future. Automatically. Because that's how their brains will have been wired. We're helping them understand what it means to manage their emotions, to control their own impulses, to consider others' feelings, to think about consequences, to make thoughtful decisions, and much more. We're helping them develop their brains and become people who are better friends, better siblings, better sons and daughters, and better human beings. Then, one day, better parents themselves.

As a huge bonus, the more we help build our kids' brains, the less we have to struggle to achieve the short-term goal of gaining cooperation. Encouraging cooperation *and* building the brain: these are the dual goals—the external and the internal—that guide a loving, effective, Whole-Brain approach to discipline. It's parenting with the brain in mind!

Accomplishing Our Goals: Saying No to the Behavior, but Yes to the Child

How do parents typically accomplish their discipline goals? Most commonly, through threats and punishment. Kids misbehave, and the immediate parental reaction is to offer consequences with both guns blazing.

Kids act, parents react, then kids react. Rinse, lather, repeat. And for many parents—probably for *most* parents—consequences (along with a healthy dose of yelling) are pretty much the primary go-to discipline strategy: time-outs, spanking, taking away a privilege, grounding, and on and on. No wonder there's so much drama! But as we'll explain, it's possible to discipline in a way that actually removes many of the reasons we give consequences in the first place.

To take it even further, consequences and punitive reactions are actually often counterproductive, not only in terms of building brains, but even when it comes to getting kids to cooperate. Based on our personal and clinical experience, as well as the latest science about the developing brain, we can tell you that automatically giving consequences is not the best way to accomplish the goals of discipline.

What is? That's the foundation of the No-Drama Discipline approach, and it comes down to one simple phrase: *connect and redirect.*

Connect and Redirect

Again, every child, like every parenting situation, is different. But one constant that's true in virtually every encounter is that the first step in effective discipline is to connect with our children emotionally. *Our relationship with our kids should be central to everything we do.* Whether we're playing with them, talking with them, laughing with them, or, yes, disciplining them, we want them to experience at a deep level the full force of our love and affection, whether we're acknowledging an act of kindness or addressing a misbehavior. Connection means that we give our kids our attention, that we respect them enough to listen to them, that we value their contribution to problem solving, and that we communicate to them that we're on their side—whether we like the way they're acting or not.

When we discipline we want to join with our kids in a deep way that demonstrates how much we love them. In fact, when our children are misbehaving, that's often when they most need connection with us. Disciplinary responses should change based on a child's age, temperament, and stage of development, along with the context of the situation. But the constant throughout the entire disciplinary interaction should be the clear communication of the deep connection between parent and child. Relationship trumps any one particular behavior.

However, *connection isn't the same thing as permissiveness.* Connecting with our kids during discipline doesn't mean letting them do whatever they want. In fact, just the opposite. Part of truly loving our kids, and giving them what they need, means offering them clear and consistent boundaries, creating predictable structure in their lives, as well as having high expectations for them. Children need to understand the way the world works: what's permissible and what's not. A well-defined understanding of rules and boundaries helps them achieve success in relationships and other areas of their lives. When they learn about structure in the safety of their home, they will be better able to flourish in outside environments—school, work, relationships—where they'll face numerous expectations for appropriate behavior. Our children need repeated experiences that allow them to develop wiring in their brain that helps them delay gratification, contain urges to react aggressively toward others, and flexibly deal with not getting their way. The absence of limits and boundaries is actually quite stressful, and stressed kids are more reactive. So when we say no and set limits for our children, we help them discover predictability and safety in an otherwise chaotic world. And we build brain connections that allow kids to handle difficulties well in the future.

In other words, *deep, empathic connection can and should be combined with clear and firm boundaries that create needed structure in children's lives.* That's where "redirect" comes in. Once we've connected with our child and helped her calm herself so she can hear us and fully understand what we're saying, we can then redirect her toward more appropriate behavior and help her see a better way to handle herself.

But keep in mind, redirection is rarely going to be successful while a child's emotions are running high. Consequences and lessons are ineffective as long as a child is upset and unable to hear the lessons you're offering. It's like trying to teach a dog to sit while he's fighting another dog. A fighting dog won't sit. But if you can help a child calm down, receptiveness will emerge and allow her to understand what you're trying to tell her, much more quickly than if you just punished or lectured her.

That's what we explain when people ask about the demands of connecting with children. Someone might say, "That sounds like a respectful and loving way to discipline, and I can see how it would help my kids in the long run, and even make discipline easier down the road. But come on! I've got a job! And other kids! And dinner to make, and piano and ballet and Little League and a hundred other things. I'm barely keeping my head above water as it is! How am I supposed to find the extra time necessary to connect and redirect when I discipline?"

We get it. We really do. Both of us work, our spouses work, and we're both committed parents. It's not easy. But what we've learned as we've practiced the principles and strategies we discuss in the following chapters is that No-Drama Discipline isn't some sort of luxury available only to people with all kinds of extra time on their hands. (We're not sure that kind of parent actually exists.) It's not that the Whole-Brain approach requires that you carve out tons of extra time to engage your kids in discussions about the right way to do things. In fact, No-Drama Discipline is all about taking ordinary, in-the-moment parenting situations and using them as opportunities to reach your kids and teach them what's important. You might think that yelling "Knock it off!" or "Quit whining!" or giving an immediate time-out would be quicker, simpler, and more effective than connecting with a child's feelings. But as we'll soon explain, paying attention to your child's emotions will usually lead to greater calm and cooperation, and do so much more quickly, than will a dramatic parental outburst that escalates the emotions all around.

And here's the best part. When we avoid bringing extra chaos and drama to disciplinary situations—in other words, when we combine clear and consistent limits with loving empathy—everyone wins. Why? For one thing a No-Drama, Whole-Brain approach makes life easier for both parents and kids. In high-stress moments—for instance, when your child threatens to throw the TV remote into the toilet mere seconds before the season finale of your beloved hospital drama—you can appeal to the higher, thinking part of her brain, rather than triggering the lower, more reactive part. (We explain this strategy in detail in Chapter 3.) As a result, you're going to be able to avoid most of the yelling and crying and anger that discipline so often causes, not to mention keeping the remote dry and getting you to your program long before the first ambulance rolls onto your screen.

More important, connecting and redirecting will, to put it as simply as possible, help your kids become better human beings, both now and as they grow toward adulthood. It will build the internal skills they'll need throughout their lives. Not only will they move from a reactive state to a receptive place where they can actually learn—that's the external, cooperation part—but connections in their brain will be built as well. These connections will allow them to grow more and more into people who know how to control themselves, think about others, regulate their emotions, and make good choices. You'll be helping them build an internal compass they can learn to rely on. Rather than simply telling them what to do and demanding that they conform to your requests, you'll be giving them experiences that strengthen their executive functions and develop skills related to empathy, personal insight, and morality. That's the internal, brain-building part.

The research is really clear on this point. Kids who achieve the best outcomes in life—emotionally, relationally, and even educationally—have parents who raise them with a high degree of connection and nurturing, while also communicating and maintaining clear limits and high expectations. Their parents remain consistent while still interacting with them in a way that communicates love, respect, and compassion. As a result, the kids are happier, do better in school, get into less trouble, and enjoy more meaningful relationships.

You won't always be able to discipline in a way that both connects and redirects. We don't do it perfectly with our own kids, either. But the more we connect and redirect, the less drama we see when we respond to our kids' misbehavior. Even better, they learn more, they build better relationship and conflict-resolution skills, and they enjoy an even stronger relationship with us as they grow and develop.

About the Book

What's involved in creating a discipline strategy that's high on relationship and low on drama? That's what the rest of the book will explain. Chapter 1, "ReTHINKING Discipline," poses some questions about what discipline is, helping you identify and develop your own discipline approach with these No-Drama strategies in mind. Chapter 2, "Your Brain on Discipline," discusses the developing brain and its role in discipline. Chapter 3, "From Tantrum to Tranquility," will focus on the "connect" aspect of discipline, emphasizing the importance of communicating that we love and embrace our children for who they are, even in the midst of a disciplinary moment. Chapter 4 stays with this theme, offering specific strategies and suggestions for connecting with kids so that they can calm down enough to really hear us and learn, thus making better decisions in both the short term and the long term.

Then it's time to redirect, which is the focus of Chapter 5. The emphasis will be on helping you remember the one definition of discipline (to teach), two key principles (wait until your child is ready, and be consistent but not rigid), and three desired outcomes (insight, empathy, and repair). Chapter 6 then concentrates on specific redirection strategies you can use for achieving the immediate goal of eliciting cooperation in the moment, and for teaching kids about personal insight, relational empathy, and taking steps toward making good choices. The book's concluding chapter offers four messages of hope intended to help you take the pressure off yourself as you discipline. As we'll explain, we all mess up when we discipline. We're all human. There is no such thing as a "perfect parent." But if we model how to deal with our mistakes and then repair the relationship afterward, even our less-than-perfect responses to misbehavior can be valuable and give kids opportunities to deal with difficult situations and therefore develop new skills. (Phew!) No-Drama Discipline isn't about perfection. It's about personal connection and repairing ruptures when they inevitably occur.

You'll see that we've included a "Further Resources" section at the back of the book. We hope this additional material will add to your experience of reading the book and help you implement the "connect and redirect" strategies in your own home. The first document we call a "Refrigerator Sheet." It contains the most essential concepts from the book, presented so you can easily remind yourself of the core No-Drama principles and strategies. Feel free to copy this sheet and hang it on your refrigerator, tape it to the dashboard of your car, or post it anywhere else it might be helpful.

Next you'll see a section called "When a Parenting Expert Loses It," which tells stories of when we, Dan and Tina, have flipped our lids and taken the low road in our own roles as parents, rather than disciplining from a No-Drama, Whole-Brain approach. In sharing these stories with you, we simply want to acknowledge that none of us is perfect, and that all of us make mistakes with our kids. We hope you'll laugh along with us as you read, and not judge us too harshly.

Next comes "A Note to Our Child's Caregivers." These pages are just what you'd expect: a note you can give to the other people who take care of your children. Most of us rely on grandparents, babysitters, friends, and others to help us raise our kids. This note lays out a brief and simple list of the key No-Drama principles. It's similar to the Refrigerator Sheet, but it's written for someone who has *not* read *No-Drama Discipline*. That way you don't have to ask your in-laws to buy and read the entire book (although nobody's stopping you from doing that if you want!).

After the note to caregivers, you'll see a list called "Twenty Discipline Mistakes Even Great Parents Make." This is one more set of reminders to help you think through the principles and issues we raise in the coming chapters. The book then closes with an excerpt from our earlier book, *The Whole-Brain Child*. By reading through this excerpt, you can get a better idea of what we mean when we talk about parenting from a Whole-Brain perspective. It's not necessary that you read this excerpt to understand what we present here, but it's there if you'd like to go deeper into these ideas and learn other concepts and strategies for building your children's brains and leading them toward health, happiness, and resilience.

Our overall goal in this book is to deliver a message of hope that will transform how people understand and practice discipline. One of the typically least pleasant parts of working with children—discipline—can actually be one of the most meaningful, and it doesn't have to be full of constant drama and reactivity for both you and your child. Children's misbehavior really can be transformed into better connections both in your relationship and within your child's brain. Disciplining from a Whole-Brain perspective will allow you to completely shift the way you think about your interactions with your children when they misbehave, and recognize those moments as opportunities to build skills that will help them now and into adulthood, not to mention making life easier and more enjoyable for everyone in the family.

No-Drama
Discipline

CHAPTER 1

ReTHINKING Discipline

Here are some actual statements we've heard from parents we've worked with. Do any of them resonate with you?

Do these comments sound familiar? So many parents feel like this. They want to handle things well when their kids are struggling to do the right thing, but more often than not, they end up simply *reacting* to a situation, rather than working from a clear set of principles and strategies. They shift into autopilot and give up control of their more intentional parenting decisions.

Autopilot may be a great tool when you're flying a plane. Just flip the switch, sit back and relax, and let the computer take you where it's been preprogrammed to go. But when it comes to disciplining children, working from a preprogrammed autopilot isn't so great. It can fly us straight into whatever dark and stormy cloud bank is looming, meaning parents and kids alike are in for a bumpy ride.

Instead of being reactive, we want to be responsive to our kids. We want to be *intentional* and make conscious decisions based on principles we've thought about and agreed on beforehand. Being intentional means considering various options and then choosing the one that engages a thoughtful approach toward our intended outcomes. For No-Drama Discipline, this means the short-term external outcome of behavioral boundaries and structure and the long-term internal outcome of teaching life skills.

Let's say, for example, your four-year-old hits you. Maybe he's angry because you told him you needed to finish an email before you could play Legos with him, and he responded by slapping you on the back. (It's always surprising, isn't it, that a person that small can inflict so much pain?)

What do you do? If you're on autopilot, not working from a specific philosophy for how to handle misbehavior, you might simply react immediately without much reflection or intention. Maybe you'd grab him, possibly harder than you should, and tell him through clenched teeth, "Hitting is not OK!" Then you might give him some sort of consequence, maybe marching him to his room for a time-out.

Is this the worst possible parental reaction? No, it's not. But could it be better? Definitely. *What's needed is a clear understanding of what you actually want to accomplish when your child misbehaves.*

That's the overall goal of this chapter, to help you understand the importance of working from an intentional philosophy and having a clear and consistent strategy for responding to misbehavior. As we said in the introduction, the dual goals of discipline are to promote good external behavior in the short term and build the internal structure of the brain for better behavior and relationship skills in the long term. Keep in mind that discipline is ultimately about teaching. So when you clench your teeth, spit out a rule, and give a consequence, is that going to be effective in teaching your child about hitting?

Well, yes and no. It *might* achieve the short-term effect of getting him not to hit you. Fear and punishment can be effective in the moment, but they don't work over the long term. And are fear, punishment, and drama really what we want to use as primary motivators of our children? If so, we teach that power and control are the best tools to get others to do what we want them to do.

Again, it's completely normal to just react when we get angry, especially when someone inflicts physical or emotional pain on us. But there are better responses, responses that can achieve the same short-term goal of reducing the likelihood of the unwanted behavior in the future, while also building skills. So rather than just fearing your response and inhibiting an impulse in the future, your child will undergo a learning experience that creates an internal skill beyond simply an association of fear. And all of this learning can occur while reducing the drama in the interaction and strengthening your connection with your child.

Let's talk about how you can respond to make discipline less of a fear-creating reaction and more of a skill-building response on your part.

The Three Questions: Why? What? How?

Before you respond to misbehavior, take a moment to ask yourself three simple questions:

1. *Why did my child act this way?* In our anger, our answer might be "Because he's a spoiled brat" or "Because he's trying to push my buttons!" But when we approach with curiosity instead of assumptions, looking deeper at what's going on behind a particular misbehavior, we can often understand that our child was trying to express or attempt something but simply didn't handle it appropriately. If we understand this, we ourselves can respond more effectively—and compassionately.

2. *What lesson do I want to teach in this moment?* Again, the goal of discipline isn't to give a consequence. We want to teach a lesson—whether it's about self-control, the importance of sharing, acting responsibly, or anything else.

3. *How can I best teach this lesson?* Considering a child's age and developmental stage, along with the context of the situation (did he realize the bullhorn was switched on before he raised it to the dog's ear?), how can we most effectively communicate what we want to get across? Too often, we respond to misbehavior as if consequences were the goal of discipline. Sometimes natural con-

sequences result from a child's decision, and the lesson is taught without our needing to do much. But there are usually more effective and loving ways to help our kids understand what we're trying to communicate than to immediately hand out one-size-fits-all consequences.

By asking ourselves these three questions—why, what, and how—when our children do something we don't like, we can more easily shift out of autopilot mode. That means we'll be much more likely to respond in a way that's effective in stopping the behavior in the short term while also teaching bigger, long-lasting life lessons and skills that build character and prepare kids for making good decisions in the future.

Let's look more closely at how these three questions might help us respond to the four-year-old who slaps you while you're emailing. When you hear the smack and feel the tiny, hand-shaped imprint of pain on your back, it may take you a moment to calm down and avoid simply reacting. It's not always easy, is it? In fact, our brains are programmed to interpret physical pain as a threat, which activates the neural circuitry that can make us more reactive and put us in a "fight" mode. So it takes some effort, sometimes intense effort, to maintain control and practice No-Drama Discipline. We have to override our primitive reactive brain when this happens. Not easy. (By the way, this gets much harder to do if we're sleep deprived, hungry, overwhelmed, or not prioritizing self-care.) This pause between reactive and responsive is the beginning of choice, intention, and skillfulness as a parent.

So as quickly as possible, you want to try to pause and ask yourself the three questions. Then you can see much more clearly what's going on in your interaction with your child. Every situation is different and depends on many different factors, but the answers to the questions might look something like this:

1. *Why did my child act this way?* He hit you because he wanted your attention and wasn't getting it. Sounds pretty typical for a four-

year-old, doesn't it? Desirable? No. Developmentally appropriate? Absolutely. It's hard for a child this age to wait, and big feelings surfaced, making it even harder. He's not yet old enough to consistently calm himself effectively or quickly enough to prevent acting out. You wish he'd just soothe himself and with composure declare, "Mom, I'm feeling frustrated that you're asking me to keep waiting, and I'm having a strong, aggressive impulse to hit you right now—but I have chosen not to and am using my words instead." But that's not going to happen. (It would be pretty funny if it did.) In that moment, hitting is your son's default strategy for expressing his big feelings of frustration and impatience, and he needs some time and skill-building practice to learn how to handle both delaying gratification and appropriately managing anger. *That's* why he hit you.

That feels much less personal, doesn't it? *Our kids don't usually lash out at us because they're simply rude, or because we're failures as parents. They usually lash out because they don't yet have the capacity to regulate their emotional states and control their impulses.* And they feel safe enough with us to know that they won't lose our love, even when they're at their worst. In fact, when a four-year-old doesn't hit and acts "perfect" all the time, we have concerns about the child's bond with his parent. When children are securely attached to their parents, they feel safe enough to test that relationship. In other words, your child's misbehavior is often a sign of his trust and safety with you. Many parents notice that their children "save it all up for them," behaving much better at school or with other adults than they do at home. This is why. These flare-ups are often signs of safety and trust, rather than just some form of rebellion.

2. *What lesson do I want to teach in this moment?* The lesson is *not* that misbehavior merits a consequence, but that there are better ways of getting your attention and managing his anger than resorting to violence. You want him to learn that hitting isn't OK, and that there are lots of *appropriate* ways to express his big feelings.

3. *How can I best teach this lesson?* While giving him a time-out or some other unrelated consequence might or might not make your son think twice next time about hitting, there's a better alternative. What if you connected with him by pulling him to you and letting him know he has your full attention? Then you could acknowledge his feelings and model how to communicate those emotions: "It's hard to wait. You really want me to play, and you're mad that I'm at the computer. Is that right?" Most likely you'll receive an angry "Yes!" in response. That's not a bad thing; he'll know he has your attention. And you'll have his, too. You can now talk with him and, as he becomes calmer and better able to listen, get eye contact, explain that hitting is never all right, and talk about some alternatives he could choose—like using his words to express his frustration—the next time he wants your attention.

INSTEAD OF JUST REACTING...

This approach works with older kids as well. Let's look at one of the most common issues faced by parents everywhere: homework

ASK THE THREE QUESTIONS

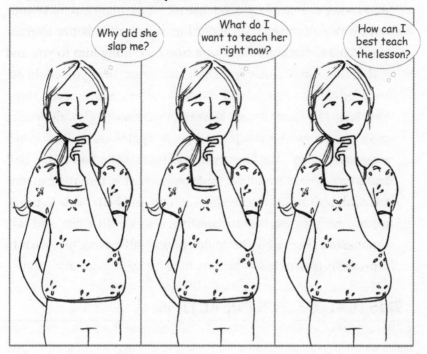

battles. Imagine that your nine-year-old is seriously struggling when it's time to study, and you two are going round and round on a regular basis. At least once a week she melts down. She gets so frustrated she ends up in tears, yelling at you and calling her teachers "mean" for assigning such difficult homework and herself "stupid" for having trouble. After these proclamations she buries her face in the crook of her arm and collapses in a puddle of tears on the table.

For a parent, this situation can be every bit as maddening as being slapped on the back by a four-year-old. An autopilot response would be to give in to the frustration and, in the heat of anger, argue with your daughter and lecture her, blaming her for managing her time poorly and not listening well enough during class. You're probably familiar with the "If you had started earlier, when I asked you to, you'd be done by now" lecture. We've never heard of a kid responding to that lecture with "You're right, Dad. I really should have started when you asked. I'll take responsibility for not beginning when I was supposed to, and I've learned my lesson. I'll just jump right on my homework earlier tomorrow. Thanks for enlightening me on this."

Instead of the lecture, what if you asked the why-what-how questions?

1. *Why did my child act this way?* Again, disciplinary approaches are going to change depending on who your child is and what her personality is like. Maybe homework is a struggle for her and she feels frustrated, like it's a battle she can never win. Maybe there's something about it that feels too hard or overwhelming and makes her feel bad about herself, or maybe she's just needing more physical activity. The main feelings here could be frustration and helplessness.

   Or maybe school isn't usually that tough for her, but she melted down because she's tired and feeling overwhelmed today. She got up early, went to school for six hours, then had a Girl Scouts meeting that lasted right up to dinnertime. Now that she's eaten, she's supposed to sit at the kitchen table and work on fractions for forty-five minutes? No wonder she's freaking out a bit. That's a lot to ask of a nine-year-old (or even an adult!). That doesn't mean she doesn't still need to do her homework, but it can change your perspective—and your response—when you realize where she's coming from.

2. *What lesson do I want to teach in this moment?* It might be that you want to teach about effective time management and responsibility. Or about making choices regarding which activities to participate in. Or about how to handle frustration more adaptively.

3. *How can I best teach this lesson?* However you answer question 2, a lecture when she's already upset definitely isn't the best approach. This isn't a teachable moment, because the emotional, reactive parts of her brain are raging, overwhelming the more calm, rational, thinking, and receptive parts of her brain. So instead, you might want to help her with her fractions and just get through

this particular crisis: "I know it's a lot tonight and you're tired. You can do this. I'll sit with you and we'll knock it out." Then once she's calmed down and you two are sharing a bowl of ice cream—or maybe even the next day—you can discuss whether she's over-scheduled, or consider that she's really struggling to understand a concept, or explore the possibility that she's talking with friends in class and bringing home unfinished classwork, meaning she ends up with more homework. Ask her questions, and problem-solve together to figure out what's going on. Ask what's getting in the way of completing her homework, why she thinks it's not working well, and what her suggestions would be. Look at the whole experience as an opportunity to collaborate on improving the homework experience. She might need some help building skills for coming up with solutions, but involve her in the process as much as possible.

Remember to pick a time when you're both in a good, recep-tive state of mind, then begin by saying something like, "The homework situation isn't working very well, is it? I bet we can find a better way. What do you think might work?" (By the way, we'll give you lots of specific, practical suggestions to help with this type of conversation in Chapter 6, where we discuss No-Drama redirection strategies.)

Different kids will require different responses to the why-what-how questions, so we're not saying that any of these specific answers will necessarily apply to your children at a given time. The point is to look at discipline in a new way, to rethink it. Then you can be guided by an overall philosophy when you interact with your kids, rather than simply reacting with whatever pops out when your kids do something you don't like. Why-what-how questions give us a new way of moving from reactive parenting to receptive and intentional Whole-Brain parenting strategies.

Granted, you won't always have time to think through the three questions. When good-natured wrestling in the living room turns

INSTEAD OF LECTURING...

ASK THE THREE QUESTIONS

into a bloody cage match, or when you have young twins who are already late for ballet, it's not that easy to go through a three-question protocol. We get it. It may sound completely unrealistic that you'd have time to be this mindful in the heat of the moment.

We're not saying you'll do it perfectly every time, or that you'll immediately be able to think through your response when your kids get upset. But the more you consider and practice this approach, the more natural and automatic it will become to offer a quick assessment and respond with an intentional response. It can even become your default, your go-to. With practice, these questions can help you remain intentional and receptive in the face of previously reaction-inducing interactions. Asking why, what, and how can help create an internal sense of clarity even in the face of external chaos.

As a result, you'll receive the bonus of having to discipline less and less, because not only will you be shaping your child's brain so that he makes better decisions and learns the connection between his feelings and his behavior, but you'll be more attuned to what's happening with him—why he does what he does—meaning that you'll be better able to guide him before things escalate. Plus, you'll be more able to see things from his perspective, which will let you recognize when he needs your help, rather than your wrath.

Can't vs. Won't: Discipline Isn't One-Size-Fits-All

To put it simply, asking the why-what-how questions helps us remember who our kids are and what they need. The questions challenge us to be conscious of the age and unique needs of each individual. After all, what works for one child may be the exact opposite of what her brother needs. And what works for one child one minute might not work for the same child ten minutes later. So don't think of discipline as a one-size-fits-all solution. Instead, *remember how important it is to discipline this one child in this one moment.*

Too often, when we discipline on autopilot, we respond to a situation much more from *our* general state of mind than from what our child needs at that particular time. It's easy to forget that our children are just that—children—and to expect behavior beyond their developmental capacity. For example, we can't expect a four-year-old to handle his emotions well when he's angry that his mom is still on the computer, any more than we can expect a nine-year-old not to freak out about homework from time to time.

Tina recently saw a mother and grandmother shopping. They had buckled a little boy, who looked about fifteen months old, into their cart. As the women browsed, looking at purses and shoes, the boy cried and cried, clearly wanting to get out of the cart. He *needed* to move and walk and explore. The caregivers absentmindedly handed him items to distract him, which just frustrated him more. This little boy couldn't talk, but his message was clear: "You're asking way too much of me! I need you to see what I need!" His behavior and emotional wails were completely understandable.

In fact, we should *assume* that kids will sometimes experience and display emotional reactivity, as well as "oppositional" behavior.

Developmentally, they're not working from fully formed brains yet (as we'll explain in Chapter 2), so they are literally incapable of meeting our expectations all of the time. That means that *when we discipline, we must always consider a child's developmental capacity, particular temperament, and emotional style, as well as the situational context.*

A valuable distinction is the idea of *can't* vs. *won't*. Parental frustration radically and drastically decreases when we distinguish between a *can't* and a *won't*. Sometimes we assume that our kids *won't* behave the way we want them to, when in reality, they simply *can't*, at least not in this particular moment.

The truth is that a huge percentage of misbehavior is more about can't than won't. The next time your child is having a hard time managing herself, ask yourself, "Does the way she's acting make sense, considering her age and the circumstances?" Much more often than not, the answer will be yes. Run errands for hours with a three-year-old in the car, and she's going to get fussy. An eleven-year-old who stayed out late watching fireworks the previous night and then had to get up early for a student council car wash the next morning is likely to melt down sometime during the day. Not because he *won't* keep it together, but because he *can't*.

We make this point to parents all the time. It was especially effective with one single father who visited Tina in her office. He was at his wits' end because his five-year-old clearly demonstrated the ability to act appropriately and make good decisions. But at times, his son would melt down over the smallest thing. Here's how Tina approached the conversation.

> I began by trying to explain to this dad that at times his son *couldn't* regulate himself, which meant that he wasn't *choosing* to be willful or defiant. The father's body language in response to my explanation was clear. He crossed his arms and leaned back in his chair. Although he didn't literally roll his eyes, it was clear he wasn't about to start a Tina Bryson fan club. So I said, "I'm getting the sense you don't agree with me here."

He responded, "It just doesn't make sense. Sometimes he's great about handling even big disappointments. Like last week when he didn't get to go to the hockey game. Then other times he'll completely lose his mind because he can't have the blue cup because it's in the dishwasher! It's not about what he can't do. He's just spoiled and needs stricter discipline. He needs to learn how to obey. And he *can*! He's already proven he can totally choose how to handle himself."

I decided to take a therapeutic risk—doing something out of the ordinary without knowing quite how it would go. I nodded, then asked, "I bet you're a loving and patient dad most of the time, right?"

He replied, "Yes, most of the time. Sometimes, of course, I'm not."

I tried to communicate some humor and playfulness in my tone as I said, "So you *can* be patient and loving, but sometimes you're choosing not to be?" Fortunately, he smiled, beginning to see where I was going. So I pressed on. "If you loved your son, wouldn't you make better choices and be a good dad *all* of the time? Why are you choosing to be impatient or reactive?" He began to nod and broke out in an even bigger smile, acknowledging my playfulness as the point sank in.

I continued. "What is it that makes it hard to be patient?"

He said, "Well, it depends on how I'm feeling, like if I'm tired or I've had a rough day at work or something."

I smiled and said, "You know where I'm going with this, don't you?"

Of course he did. Tina went on to explain that a person's capacity to handle situations well and make good decisions can really fluctuate according to the circumstances and the context of a given situation. Simply because we're human, our capacity to handle ourselves well is not stable and constant. And that's certainly the case with a five-year-old.

The father clearly understood what Tina was saying: that it's misguided to assume that just because his son could handle himself well in one moment, he'd always be able to do so. And that when his son didn't manage his feelings and behaviors, it wasn't evidence that he was spoiled and needed stricter discipline. Rather, he needed understanding and help, and through emotional connection and setting limits, the father could increase and expand his son's capacity. *The truth is that for all of us, our capacity fluctuates given our state of mind and state of body, and these states are influenced by so many factors—especially in the case of a developing brain in a developing child.*

Tina and the father talked further, and it was clear that he had fully understood Tina's point. He got the difference between can't and won't, and he saw that he was imposing rigid and developmentally inappropriate (one-size-fits-all) expectations on his young son, as well as on the boy's sister. This new perspective empowered him to switch off his parental autopilot and start working on making intentional, moment-by-moment decisions with his children, both of whom had their own particular personality and needs at different moments. *The father realized that not only could he still set clear, firm boundaries, but he could do so even more effectively and respectfully, because he was taking into account each child's individual temperament and fluctuating capacity, along with the context of each situation.* As a result, he'd be able to achieve both disciplinary goals: to see less overall uncooperativeness from his son, and to teach him important skills and life lessons that would help him as he grew into a man.

This father was learning to challenge certain assumptions in his own thinking, such as that misbehavior is always willful opposition instead of a moment of difficulty while trying to manage feelings and behaviors. Future conversations with Tina led him to question not only this assumption, but also his emphasis on having his son and daughter obey him unconditionally and without exception. Yes, he reasonably and justifiably wanted his discipline to encourage cooperation from his children. But complete and unquestioning obedience? Did he want his kids to grow up blindly obeying everyone their whole lives? Or would he rather have them develop their own individual personalities and identities, learning along the way what it means to get along with others, observe limits, make good decisions, be self-disciplined, and navigate difficult situations by thinking for themselves? Again, he got the point, and it made all the difference for his children.

One other assumption this father began to challenge within himself was that there's some silver bullet or magic wand that can be used to address any behavioral issue or concern. We wish there were such a cure-all, but there's not. It's tempting to buy into one discipline practice that promises to work all the time and in every situation or to radically change a kid in a few days. But the dynamics of interacting with children are always much more complex than that. Behavioral issues simply can't be resolved with a one-size-fits-all approach that we apply to every circumstance or environment or child.

Let's take a few minutes now and discuss the two most common one-size-fits-all disciplinary techniques that parents rely on: spanking and time-outs.

Spanking and the Brain

One autopilot response that a number of parents resort to is spanking. We often get asked where we stand on the subject.

Although we're really big advocates for boundaries and limits, we are both strongly against spanking. Physical punishment is a complex and highly charged topic, and a full discussion of the research, the various contexts in which physical punishment takes place, and the negative impacts of spanking is beyond the scope of this book. But based on our neuroscientific perspective and review of the research literature, we believe that spanking is likely to be counterproductive when it comes to building respectful relationships with our children, teaching kids the lessons we want them to learn, and encouraging optimal development. We also believe that children should have the right to be free from any form of violence, especially at the hands of the people they trust most to protect them.

We know there are all kinds of parents, all kinds of kids, and all kinds of contexts in which discipline takes place. And we certainly understand that frustration, along with the desire to do the right thing for their children, leads some parents to use spanking as a discipline strategy. But the research consistently demonstrates that even when parents are warm, loving, and nurturing, not only is spanking children less effective in changing behavior in the long run, it's associated with negative outcomes in many domains. Granted, there are plenty of non-spanking discipline approaches that can be just as damaging as spanking. Isolating children for long periods of time, humiliating them, terrifying them by screaming threats, and using other forms of verbal or psychological aggression are all examples of disciplinary practices that wound children's minds even when their parents never physically touch them.

We therefore encourage parents to avoid *any* discipline approach that is aggressive, inflicts pain, or creates fear or terror. For one thing, it's counterproductive. The child's attention shifts from her own behavior and how to modify it, to the caregiver's response to the behavior, meaning that the child no longer considers her own actions at all. Instead, she thinks only about how unfair and mean her parent was to hurt her—or even how scary her parent was in that moment. *The parental response, then, undermines both of the primary goals of discipline—changing behavior and building the brain—because it sidesteps an opportunity for the child to think about her own behavior and even feel some healthy guilt or remorse.*

Another important problem with spanking is what happens to the child physiologically and neurologically. The brain interprets pain as a threat. So when a parent inflicts physical pain on a child, that child faces an unsolvable biological paradox. On one hand, we're all born with an instinct to go toward our caregivers for protection when we're hurt or afraid. But when our caregivers are also the *source* of the pain and fear, when the parent has caused the state of terror inside the child by what he or she has done, it can be very confusing for the child's brain. One circuit drives the child to try to escape the parent who is inflicting pain; another circuit drives the child *toward* the attachment figure for safety. So when the parent is the source of fear or pain, the brain can become disorganized in its functioning, as there is no solution. We call this at the extreme a form of disorganized attachment. The stress hormone cortisol, released with such a disorganized internal state and repeated interpersonal experiences of rage and terror, can lead to long-lasting negative impacts on the brain's development, as cortisol is toxic to the brain and inhibits healthy growth. Harsh and severe punishment can actually lead to significant changes in the brain, such as the death of brain connections and even brain cells.

Another problem with spanking is that it teaches the child that the parent has no effective strategy short of inflicting bodily pain. That's a direct lesson every parent should consider quite deeply: do we want to teach our kids that the way to resolve a conflict is to inflict physical pain, particularly on someone who is defenseless and cannot fight back?

Looking through the lens of the brain and body, we know that humans are instinctually wired to avoid pain. And it is also the same part of the brain that mediates physical pain that processes social rejection. Inflicting physical pain is also creating social rejection in the child's brain. Since children can't be perfect, we see the importance of the findings indicating that while spanking often stops a behavior in a particular moment, it's not as effective at changing behavior in the long run. Instead, children will often just get better at concealing what they've done. In other words, the danger is that kids will do whatever it takes to avoid the pain of physical punishment (and social rejection), which will often mean more lying and hiding—not collaboratively communicating and being open to learning.

One final point about spanking has to do with which part of the brain we want to appeal to and develop with our discipline. As we'll explain in the next chapter, parents have the option of engaging the higher, thinking part of the child's wise brain, or the lower, more reactive, reptilian part. If you threaten or physically attack a reptile, what kind of a response do you think you'll get? Imagine a cornered cobra, spitting at you. There is nothing wise or connecting about being reactive.

When we are threatened or physically attacked, our reptilian or primitive brain takes over. We move into an adaptive survival mode, often called "fight, flight, or freeze." We can also faint, a response that occurs in some when they feel totally helpless. Likewise, when we cause our kids to experience fear, pain, and anger, we activate an increase in the flow of energy and information to the primitive, reactive brain, instead of directing the flow to the receptive, thinking, more sophisticated and potentially wise regions of the brain that allow kids to make healthy and flexible choices and handle their emotions well.

Do you want to trigger reactivity in your child's primitive brain, or engage her thinking, rational brain in being receptive and openly engaged with the world? When we activate the reactive states of the brain, we miss the chance to develop the thinking part of the brain. It's a lost opportunity. What's more, we have *so many* other, more effective options for disciplining our kids—strategies that give children practice using their "upstairs brain" so that it's stronger and further developed, meaning that they're much better able to be responsible people who do the right thing more often than not. (Much more about that in Chapters 3–6.)

What About Time-outs? Aren't They an Effective Discipline Tool?

These days, most parents who have decided they don't want to spank their kids assume that time-outs are the best available option. But are they? Do they help us achieve our discipline goals?

In general terms, we don't think so.

We know lots of loving parents who use time-outs as their primary discipline technique. But after exploring the research, talking to thousands of parents, and raising our own kids, we've come up with several main reasons we do not think that time-outs are the best discipline strategy. For one thing, when parents use time-outs, they often use them a lot, and out of anger. But parents can give children more positive and meaningful experiences that better achieve our dual goals of encouraging cooperation and building the brain. As we'll explain in more detail in the next chapter, brain connections are formed from repeated experiences. And what experience does a time-out give a child? Isolation. Even if you can offer a time-out in a loving manner, do you want your child's repeated experiences when she makes a mistake to be time by herself, which is often experienced, particularly by young children, as rejection?

Wouldn't it be better to have her experience what it means to do things the *right way*? So instead of a time-out, you might ask her to practice handling a situation differently. If she's being disrespectful in her tone or words, you can have her try it again and communicate what she's saying respectfully. If she's been mean to her brother, you might ask her to find three kind things to do for him before bedtime. That way, the repeated experience of positive behavior begins to get wired into her brain. (Again, we'll cover this in more depth in the following chapters.)

In short, time-outs often fail to accomplish their objective, which is *supposed* to be for children to calm down and reflect on their behavior. In our experience, time-outs frequently just make children angrier and more dysregulated, leaving them even less able to control themselves or think about what they've done. Plus, how often do you think kids use their time-out to reflect on their behavior? We've got news for you: the main thing kids reflect on while in time-out is how mean their parents are to have put them there.

When children are reflecting on their horrible luck to have such a mean, unfair mom or dad, they're missing out on an opportunity to build insight, empathy, and problem-solving skills. Putting them in time-out deprives them of a chance to practice being active, empathic decision makers who are empowered to figure things out. We want to give them opportunities to be problem solvers, to make good choices, and to be comforted when they are falling apart. You can do your kids a lot of good simply by asking, "What are some ideas you have to

WHAT PARENTS EXPECT FROM TIME-OUTS:

WHAT ACTUALLY HAPPENS DURING TIME-OUTS:

make it better and solve this problem?" Given the chance once they're calm, kids will usually do the right thing, and learn in the process.

In addition, too often time-outs aren't directly and logically linked to a particular behavior, which is a key to effective learning. Making a toilet-paper mountain means helping clean up. Riding a bike without a helmet means that rather than simply jumping on the bike and riding, for two weeks there will be a required safety check anytime the bicycle comes out of the garage. Leaving a bat at baseball practice means having to borrow a teammate's bat until the other one turns up. Those are connected parental responses that are clearly linked to the behavior. They aren't punitive or retaliatory in any way. They are focused on teaching kids lessons and helping them understand about making things right. Time-outs, though, often don't relate in any clear way to a child's poor decision or out-of-control reaction. As a result, they're often not as effective in terms of changing behavior.

Even when parents have good intentions, time-outs are often used inappropriately. We might *want* time-outs to give kids a chance to calm down and pull themselves together so they can move out of their internal chaos into calm and cooperation. But much of the time, parents use time-outs punitively, where the goal isn't to help the child return to his calm baseline or to learn an important lesson, but to punish him for some misbehavior. The calming, teaching aspect of the time-out gets totally lost.

But the biggest reason we question the value of time-outs has to do with a child's profound need for connection. Often, misbehavior is a result of a child getting overtaxed emotionally, so the expression of a need or a big feeling comes out in ways that are aggressive, disrespectful, or uncooperative. She may be hungry or tired, or maybe there's some other reason she's incapable in that moment of self-control and making a good decision. Maybe the explanation is simply that she's three, and her brain isn't sophisticated enough to understand and calmly express her feelings. So instead of doing her best to convey her crushing disappointment and anger that there's no grape juice left, she begins throwing toys at you.

INSTEAD OF A ONE-SIZE-FITS-ALL TIME-OUT...

GIVE THEM PRACTICE MAKING GOOD CHOICES

It's during these times that a child most needs our comfort and calm presence. Forcing her to go off and sit by herself can feel like abandonment to a child, especially if she's feeling out of control already. It may even send the subtle message that when she isn't "doing the right thing" you don't want to be near her. *You don't want to send the message that you'll be in relationship with her when she's "good" or happy, but you'll withhold your love and affection when she's not.* Would you want to stay in that kind of a relationship? Wouldn't we suggest to our teenagers that they might think about avoiding friends or partners who treat them like that when they've made a mistake?

IS THIS THE MESSAGE YOU WANT TO GIVE YOUR CHILD?

We're not saying that short time-outs are the worst possible discipline technique, that they cause trauma, or that there's never an instance to use them. If done appropriately with loving connection, such as sitting *with* the child and talking or comforting—what can be called a "time-in"—some time to calm down can be helpful for children. In fact, teaching kids how to pause and take some inner reflection time, some time-in, is essential for building executive functions that reduce impulsivity and harness the power of focused attention. But such reflection is created in relationship, not in complete isolation, especially for younger children. As children get older, they can benefit from inner reflection, from time-in, to focus their attention on their inner world. This is how they learn to "see the sea inside" and develop the skill to calm the storms inside. Such time-in is the basis of mindsight, of seeing one's own mind and the mind of others with insight and empathy. And mindsight includes the process of integration that enables inner states to be changed, to move from chaos or rigidity to an inner state of harmony and flexibility. Mindsight— insight, empathy, and integration—is the basis of social and emotional intelligence, so using time-in to develop inner reflective skills is how we help children and adolescents build the circuitry of such important abilities. No-Drama Discipline would use a time-in to stop behavior (the first goal) and to invite inner reflection that builds executive skills (our second goal).

One proactive strategy that can be effective is to help the child create a "calm zone" with toys, books, or a favorite stuffed animal, which she visits when she needs the time and place to calm down. That's internal self-regulation, a fundamental skill of executive function. (This is a good idea for parents, too! Maybe some chocolate, magazines, music, red wine . . .) It's not about punishment or making a child pay for her mistake. It's about offering a choice and a place that helps the child self-regulate and down-regulate, which involves downshifting out of her emotional overload.

As you'll see in the coming pages, there are dozens of other, more nurturing, relationship-building, and *effective* ways to respond to kids than to automatically give them a time-out as a one-size-fits-all default consequence for any misbehavior. The same goes for spanking, and even for giving consequences in general. Fortunately, as we'll soon explain, there are better alternatives than to spank, give a time-out, or automatically take away a toy or a privilege. Alternatives that are logically and naturally linked to the child's behavior, that build the brain, and that maintain a strong connection between parent and child.

What Is Your Discipline Philosophy?

The main point we've communicated in this chapter is that parents need to be intentional about how they respond when their kids misbehave. Rather than dramatically or emotionally reacting, or responding to every infraction with a one-size-fits-all strategy that ignores the context of the situation or a child's developmental stage, parents can work from principles and strategies that both match their belief system and respect their children as the individuals they are. No-Drama Discipline focuses not only on addressing immediate circumstances and short-term behavior, but also on building skills and creating connections in the brain that, in the long run, will help children make thoughtful choices and handle their emotions well automatically, meaning that discipline will be needed less and less.

How are you doing on this? How intentional are you when you discipline your children?

Take a moment right now and think about your normal response to your kids' behavior. Do you automatically spank, give a time-out, or yell? Do you have some other immediate go-to for when your kids act out? Maybe you simply do what your parents did—or just the opposite. The real question is, how much of your disciplinary strategy comes from an intentional and consistent approach, as opposed to simply reacting or relying on old habits and default mechanisms?

Here are some questions to ask yourself as you think about your overall discipline philosophy:

1. *Do I have a discipline philosophy?* How purposeful and consistent am I when I don't like how my kids are behaving?
2. *Is what I'm doing working?* Does my approach allow me to teach my kids the lessons I want to teach, in terms of both immediate

behavior and how they grow and develop as human beings? And am I finding that I need to address behaviors less and less, or am I having to discipline about the same behaviors over and over?

3. *Do I feel good about what I'm doing?* Does my discipline approach help me enjoy my relationship with my children more? Do I usually reflect on discipline moments and feel pleased with how I handled myself? Do I frequently wonder if there's a better way?

4. *Do my kids feel good about it?* Discipline is rarely going to be popular, but do my children understand my approach and feel my love? Am I communicating and modeling respect in a way that allows them to still feel good about themselves?

5. *Do I feel good about the messages I'm communicating to my children?* Are there times I teach lessons I don't want them to internalize—for example, that obeying what I say is more important than learning to make good decisions about doing the right thing? Or that power and control are the best ways to get people to do what we want? Or that I only want to be around them if they're pleasant?

6. *How much does my approach resemble that of my own parents?* How did my parents discipline me? Can I remember a specific experience of discipline and how it made me feel? Am I just repeating old patterns? Rebelling against them?

7. *Does my approach ever lead to my kids apologizing in a sincere manner?* Even though this might not happen on a regular basis, does my approach at least leave the door open for it?

8. *Does it allow for me to take responsibility and apologize for my own actions?* How open am I with my kids about the fact that I make mistakes? Am I willing to model for them what it means to own up to one's errors?

How do you feel right now, having asked yourself these questions? Many parents experience regret, guilt, shame, or even hopelessness when they acknowledge what has not been working and worry that they may not have been doing the best they can. But *the truth is, you have done the best you can. If you could have done better, you would have.* As you learn new principles and strategies, the goal is not to berate yourself for missed opportunities, but to try to create new opportunities. When we know better, we do better. There are things we, as experts, have learned over the years that we wish we'd known or thought about when our children were babies. Our children's brains are extremely plastic—they change their structure in response to experience—and our children can respond very quickly and very productively to new experiences. The more compassion you can have for yourself, the more compassion you can have for your child. Even the best parents realize that there will always be times they can be more intentional, effective, and respectful regarding how they discipline their children.

In the remaining chapters, our goal is to help you think about what you want for your kids when it comes to guiding and teaching them. None of us will ever be perfect. But we can take steps toward modeling calm and self-control when our kids mess up. We can ask the why-what-how questions. We can steer clear of one-size-fits-all disciplinary techniques. *We can offer the two goals of shaping external behaviors and learning internal skills.* And we can work on reducing the number of times we simply *react* (or overreact) to a situation, and increasing the times we *respond* out of a clear and receptive sense of what we believe our kids need—in each particular moment, and as they move through childhood toward adolescence and adulthood.

CHAPTER 2

Your Brain on Discipline

iz's morning was going along fine. Both of her kids had eaten breakfast, everyone was dressed, and she and her husband, Tim, were heading out the door to take their daughters to their respective schools. Then all of a sudden, when Liz uttered the most seemingly trivial statement as she closed the front door behind her—"Nina, you get in Daddy's car, and Vera, you get in the van"—everything fell apart.

Tim and their seven-year-old, Vera, had already started toward the driveway, and Liz was locking the front door when a feral scream from just behind her made her heart stop. She quickly turned around to see Nina, her four-year-old, standing on the bottom step of the porch, screaming "No!" in an astonishingly earsplitting register.

Liz looked at Tim, then at Vera, both of whom shrugged, eyes wide with confusion. Nina's long, sustained "No!" had been replaced by a staccato "No! No! No!" repeated, again, at full volume. Liz quickly knelt and pulled Nina to her, her daughter's shrieks mercifully petering out and turning into sobs.

"Honey, what is it?" Liz asked. She was dumbfounded at this outburst. "What is it?"

Nina continued to cry but was able to utter, "You took Vera yesterday!"

Liz again looked at Tim, who had walked toward them and offered a puzzled "I have no idea" shrug. Liz, her ears still ringing, tried to explain: "I know, sweetheart. That's because Vera's school is right by my work."

Nina pulled back from her mother and screamed, "But it's *my turn!*"

Now that she knew her daughter wasn't in danger, Liz took a deep breath and briefly wondered what decibel level a high-pitched scream would have to reach to actually break glass.

Vera, typically unsympathetic when it came to her sister's distress, impatiently announced, "Mom, I'm gonna be late."

Before we describe how Liz handled this classic parenting situation, we need to introduce a few simple facts about the human brain and the way it can impact our disciplinary decisions when our kids misbehave or, as in this case, just lose control of themselves. Let's begin with three foundational discoveries about the brain—we'll call them the three "Brain C's"—that can be immensely beneficial when it comes to helping you discipline effectively and with less drama, all while teaching your children important lessons about self-control and relationships.

"Brain C" #1: The Brain Is *Changing*

The first Brain C—that the brain is changing—sounds simple, but its implications are enormous and should inform just about everything we do with our kids, including discipline.

A child's brain is like a house that's under construction. The downstairs brain is made up of the brainstem and the limbic region, which together form the lower sections of the brain, often called the "reptilian brain" and the "old mammalian brain." These lower regions exist inside your skull from about the level of the bridge of your nose down to the top of your neck, and some of it, the brainstem, is well-developed at birth. We consider this downstairs brain to be much more primitive, because it's responsible for our most fundamental neural and mental operations: strong emotions; instincts like protecting our young; and basic functions like breathing, regulating sleep and wake cycles, and digestion. The downstairs brain is what causes a toddler to throw a toy or bite someone when he doesn't get his way. It can be the source of our reactivity, and its motto is a rushed "Fire! Ready! Aim!"—and often it skips the "ready" and "aim" parts altogether. It was Nina's downstairs brain that took over when she was told her mom wouldn't be driving her to school.

As you well know if you're a parent, the downstairs brain, with all of its primitive functions, is alive and well in even the youngest children. The upstairs brain, though, which is responsible for more sophisticated and complex thinking, is undeveloped at birth and begins to grow during infancy and childhood. The upstairs brain is made up of the cerebral cortex, which is the outermost layer of the brain, and it resides directly behind your forehead and continues to the back of your head like a half dome covering the downstairs brain below it. Sometimes people refer to the cortex as the "outer bark of the brain." Unlike the primitive downstairs brain, with all of its rudimentary

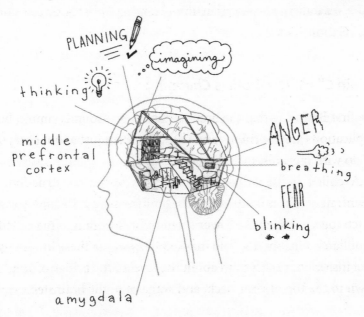

functions, the upstairs brain is responsible for a laundry list of thinking, emotional, and relational skills that allow us to live balanced, meaningful lives and enjoy healthy relationships:

- Sound decision making and planning
- Regulation of emotions and body
- Personal insight
- Flexibility and adaptability
- Empathy
- Morality

These are the very qualities we want to help instill in our children, and they all require a well-developed upstairs brain.

The problem is that the upstairs brain takes time to develop. A long time. We're sorry to report—especially if today happens to be the third time this week that your twelve-year-old left his homework binder in his locker—that the upstairs brain actually won't be fully formed until a person reaches his mid-twenties. That doesn't mean there's nothing for it to do along the way—it simply means that while the child's brain is being constructed, the adolescent brain is in a period of remodeling itself and will be changing the basic upstairs brain structures that were created in the first dozen years of life. Dan explores all of this in his book for and about adolescents called *Brainstorm*. The great news is that knowing about the brain—for you and your child or adolescent—can change the way each of you approaches learning and behaving. When we know about the brain, we can guide our minds—how we pay attention, how we think, how we feel, how we interact with others—in ways that support solid, healthy brain development across the life span.

Still, what this all means is that as much as we'd like for our kids to consistently behave as if they were fully developed, conscientious adults, with reliably functioning logic, emotional balance, and morality, they just can't yet when they are young. At least not all the time. As a result, we have to proceed accordingly and adjust our expectations. We want to turn to our nine-year-old and ask, as we comfort our five-year-old whose eye has been struck by a Nerf bullet fired at infuriatingly close range, "What were you *thinking*?"

His answer, of course, will be "I don't know" or "I *wasn't* thinking." And most likely he'll be right. His upstairs brain wasn't engaged when he aimed at his sister's pupil, just as her upstairs brain wasn't engaged yesterday when she demanded that her cousin's beach party be moved inside because she got a cut on her heel and didn't want to get sand in it. The bottom line is that no matter how smart, responsible, or conscientious your child is, it's unfair to expect her to always handle herself well, or to always distinguish between a good choice and a bad one. That's even impossible for adults to do all the time.

A good example of this gradual development can be found in a particular area of the upstairs brain called the right temporal parietal junction (TPJ).

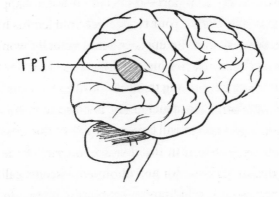

The right TPJ plays a special role when it comes to helping us understand what's going on in the mind of another. When we view a situation or a problem as someone else would, the right TPJ becomes active and works with areas in the prefrontal cortex, just behind the forehead, essentially to allow us to empathize with another. These and other areas are part of what is called a "mentalizing circuit" because they are involved in mindsight—that is, seeing the mind of others, and even of ourselves! We can build mindsight in our children as we guide them toward insight, empathy, and moral thinking. Empathy, of course, affects our moral and relational lives in significant and foundational ways. We're willing to cut someone some slack if she meant well when she messed up. We're willing to give someone the benefit of the doubt if we trust his motives.

A child, though, who is still developing and whose upstairs brain—which includes his right TPJ and prefrontal regions—is still under construction, will often be unable to consider motives and intention when he looks at a situation or problem. Ethical decisions will be much more black and white, and concerns about issues like justice and fairness will be much more clear-cut. Nina, for example, had no interest in discussing contextual information about how close her sister's school was to her mom's job. That logical, factual bit of data was irrelevant to her. She cared only that her sister had ridden with her mother yesterday, and fairness would dictate that Nina should get to ride with her today. So for Liz to understand her daughter's point of view, she needed to realize that Nina was viewing events through the lens of her still-growing upstairs brain, which wasn't always able to consider situational and contextual information.

As we'll explain in subsequent chapters, when we use our own mindsight circuits to sense the mind behind our children's behavior, we model for them how to sense the mind within themselves and others. Mindsight is a teachable skill at the heart of being empathic and insightful, moral and compassionate. Mindsight is the basis of social and emotional intelligence, and we can model this for our children as we help guide the development of their changing brains.

The point is that when we parent, and especially when we discipline, we need to work hard to understand our children's points of view, their developmental stage, and what they are ultimately capable of. This is how we use our own mindsight skills to see the mind behind our children's behavior. We don't simply react to their external actions, we tune in to the mind behind the behavior. We also must remember that what they're capable of isn't always the same; their capacity changes when they are feeling tired, hungry, or overwhelmed. Comprehending this particular Brain C, that the brain is changing and still developing, can move us to a place where we can listen to our kids with more understanding and compassion, and more fully understand why it is that they're upset and having a hard time managing themselves. It is simply unfair to assume that our children are making decisions using fully formed, perfectly functioning brains and can view the world as we do.

Think about the list of functions the upstairs brain is responsible for. Is that a realistic description of any child's character? *Of course* we'd love to see our kids demonstrate these qualities each and every moment of their lives. Who wouldn't want a child who plans ahead and consistently makes good decisions, controls his emotions and body, displays flexibility and empathy and self-understanding, and acts out of a well-developed sense of morality? But it's just not going to happen. At least not all the time. Depending on the child and the age, maybe not even frequently.

So is this an excuse for bad behavior? Do we need to simply turn a blind eye when our kids misbehave? Certainly not. In fact, a child's developing brain is simply another reason we need to set clear boundaries and help her understand what's acceptable. The fact that she doesn't have a consistently working upstairs brain, which provides *internal* constraints that govern her behavior, means that she needs to be provided with *external* constraints. And guess where those external constraints need to come from: her parents and other caregivers, and the guidelines and expectations they communicate to her. *We need to help develop our children's upstairs brain—along with all of the skills it makes possible—and while doing so, we may need to act as an external upstairs brain along the way, working with them and helping them make decisions they're not quite capable yet of making for themselves.*

We'll soon go into this idea in much greater depth, and offer practical suggestions for making it happen. For now, though, just keep this initial Brain C in mind: a child's brain is changing and developing, so we need to temper our expectations and understand that emotional and behavioral challenges are simply par for the course. *Of course* we should still teach and expect respectful behavior. But in doing so, we need to always keep in mind the changing, developing brain. Once we understand and accept this fundamental reality, we'll be much more capable of responding in a way that honors the child and the relationship, while still attending to any behaviors we need to address.

"Brain C" #2: The Brain Is *Changeable*

The second Brain C is immensely exciting and offers hope to parents everywhere: the brain is not only changing—it develops over time— but changeable—it can be molded intentionally by experience. If you read much about the brain these days, you'll likely come across the concept of "neuroplasticity," which refers to the way the brain physically changes based on experiences we undergo. As scientists put it, the brain is plastic, or moldable. Yes, the actual physical architecture of the brain changes based on what happens to us.

You may have heard about scientific studies that demonstrate evidence of neuroplasticity. In *The Whole-Brain Child,* we talk about research showing enlarged auditory centers in the brains of animals who depend on their hearing for hunting, and studies showing that for violinists, the regions of the cortex that represent the left hand— which fingers the instrument's strings at amazing speeds—are larger than normal.

Other recent studies demonstrate that children who are taught to read music and play the keyboard undergo significant changes in their brain and have an advanced capacity for what's called "spatial sensorimotor mapping." In other words, when kids learn even the fundamentals of playing piano, their brains develop differently from the brains of kids who don't, so they can more fully understand their own bodies in relationship to the objects around them. We've seen similar results in studies on people who meditate. Mindfulness exercises produce literal changes in the brain's connections, significantly affecting how well a person interacts with other people and adapts to difficult situations.

Obviously, this isn't to say that all children should take piano les-

EXPERIENCE LITERALLY CHANGES THE BRAIN

sons, or that everyone should meditate (although we wouldn't discourage either activity!). The point is that the experience of taking the lessons, like the experience of participating in mindfulness practices (or playing the violin or even practicing karate), fundamentally *and physically* changes the plastic brain—especially while it's developing in childhood and adolescence, but even throughout our lives. To take a more extreme example, early childhood abuse can leave people vulnerable to mental illness later in life. Recent studies have used functional magnetic resonance imaging (fMRI), or brain scans, to discover specific changes in certain areas of what's called the hippocampus in the brains of young adults who have experienced abuse.

They experience higher rates of depression, addiction, and post-traumatic stress disorder (PTSD). Their brains have fundamentally changed in response to the trauma they faced as children.

Neuroplasticity has enormous ramifications for what we do as parents. *If repeated experiences actually change the physical architecture of the brain, then it becomes paramount that we be intentional about the experiences we give our children.* Think about the ways you interact with your kids. How do you communicate with them? How do you help them reflect on their actions and behavior? What do you teach them about relationships—about respect, trust, and effort? What opportunities do you expose them to? What important people do you introduce into their lives? Everything they see, hear, feel, touch, or even smell impacts their brain and thus influences the way they view and interact with their world—including their family, neighbors, strangers, friends, classmates, and even themselves.

All of this takes place at the cellular level, in our neurons and in the connections among our brain cells called synapses. One way neuroscientists have expressed the idea is that "neurons that fire together wire together."

This phrase, known as "Hebb's axiom," named after the Canadian neuropsychologist Donald Hebb, essentially explains that when neurons fire simultaneously in response to an experience, those neurons

NEURONS THAT
FIRE TOGETHER
WIRE TOGETHER

become connected to each other, forming a network. And when an experience is repeated over and over, it deepens and strengthens the connections among those neurons. So when they fire together, they wire together.

The famous physiologist Ivan Pavlov was coming to terms with this idea when he found that his dogs would salivate not only when actual food appeared before them, but also when he rang the dinner bell for them to come eat. The dogs' "salivation neurons" became wired, or functionally linked, to their "dinner-bell neurons." A more recent example from the animal world appears every time the San Francisco Giants play a night game at AT&T Park. Near the end of each game swarms of seagulls appear, ready for a feast of left-behind hot dogs, peanuts, and Cracker Jack once the bayside stadium empties. Biologists are stumped as to how, exactly, the birds time their arrival for the ninth inning. Is it the increased noise of the crowd? The lure of the stadium lights? The organ playing "Take Me Out to the Ball Game" during the seventh-inning stretch? One thing seems clear, though: the birds have been conditioned, or primed, to expect food once the game ends. Neurons have fired together and subsequently been wired together.

Hebb's axiom is what causes a toddler to raise his hands and say "Hold you?" when he wants to be picked up. He hardly understands the meaning of the exact words, and obviously hasn't quite figured out his pronouns. But he knows that when he's been asked, "Do you want me to hold you?" he's been picked up. So when he wants to be held, he asks, "Hold you?" Firing and wiring.

Having neurons wire together can be a good thing. A positive experience with a math teacher can lead to neural connections that link math with pleasure, accomplishment, and feeling good about yourself as a student. But the opposite is equally true. Negative experiences with a harsh instructor or a timed test and the anxiety that accompanies it can form connections in the brain that create a serious obstacle to the enjoyment not only of math and numbers, but exams and even school in general.

The point is simple but crucial to understand: experiences lead to changes in the architecture of the brain. Practically, then, we want to keep neuroplasticity in mind when we make decisions about how we interact with kids and how they spend their time. We want to consider what neural connections are being formed and how they will play out in the future.

For example, what movies do you want your kids to see, and what activities do you want them to spend hours of time enjoying? Knowing that the plastic brain will be altered with experience, we might be less comfortable with hours spent watching certain television programs or playing violent video games. We might instead encourage our kids to engage in activities that build their capacity for relationships and for understanding other people—whether that means hanging out with friends,.playing games with their family, or participating in sports and other group activities that ask them to work with others as a team. We might even purposely create time for boredom on a summer day, so they have to go to the garage and see what interesting fun they can have with a pulley, some rope, and a roll of duct tape. (If someone comes back inside to Google the phrase "duct tape parachute for baby brother," you might want to break out the Monopoly board.)

We cannot, nor would we want to, protect or rescue our kids from all adversity and negative experiences. These challenging experiences are an important part of growing up and developing resilience, along with acquiring internal skills needed to cope with stress and failure and to respond with flexibility. What we can do is help our children make sense of their experiences so that those challenges will more likely be encoded in the brain consciously as "learning experiences," rather than unconscious associations or even traumas that limit them in the future. When parents discuss experiences and memories with their kids, the children tend to have better access to memories of those experiences. Kids whose parents talk to them about their feelings also develop a more robust emotional intelligence and can therefore be better at noticing and understanding their own and other people's feelings. Neurons that fire together wire together, changing the changeable brain.

It all comes back to the point that the brain changes in response to experience. What do you want your children to experience that will affect their changeable brains? What brain connections do you want to nurture? And more to the point in this book: knowing that a child's brain is changeable, how do you want to respond to misbehavior? After all, your kids' repeated experiences with discipline will be wiring their brains as well.

"Brain C" #3: The Brain Is *Complex*

So the brain is changing and changeable. It's also complex, which is our third Brain C. The brain is multifaceted, with different parts responsible for different tasks. Some parts are responsible for memory, others for language, others for empathy, and so on.

This third Brain C is one of the most important realities to keep in mind when it comes to discipline. The brain's complexity means that when our kids are upset, or when they're acting in ways we don't like, we can appeal to different "parts" of their brains, to different regions and ways the brain functions, with different parental responses activating different circuitry. We can therefore appeal to one part of the brain to get one result, another part to get a different result.

For example, let's go back to the upstairs and downstairs brains. If your child is melting down and out of control, which part of the brain would you rather appeal to? The one that's primitive and reactive? Or the one that's sophisticated and capable of logic, compassion, and self-understanding? Do we try to connect to the one that responds as a reptile would—with defensiveness and attacks—or to the one with the potential to calm down, problem-solve, and even apologize? The answer is obvious. We want to engage the upstairs brain's receptivity, rather than trigger the downstairs brain's reactivity. Then the higher parts of the brain can communicate and help override the lower, more impulsive and reactive parts.

When we discipline with threats—whether explicitly through our words or implicitly through scary nonverbals like our tone, posture,

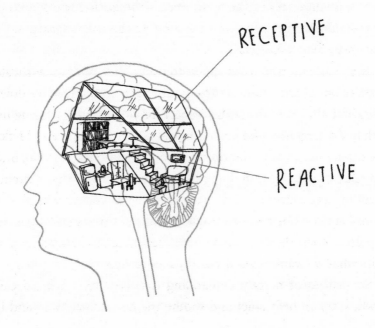

RECEPTIVE

REACTIVE

and facial expressions—we activate the defensive circuits of our child's reactive reptilian downstairs brain. We call this "poking the lizard," and we don't recommend it because it almost always leads to escalating emotions, for both parent and child. When your five-year-old throws a fit at the grocery store, and you tower over him and point your finger and insist through clenched teeth that he "calm down this instant," you're poking the lizard. You're triggering a downstairs reaction, which is almost never going to lead anywhere productive for anyone involved. Your child's sensory system takes in your body language and words and detects *threat*, which biologically sets off the neural circuitry that allows him to survive a threat from his environment—to fight, to flee, to freeze, or to faint. His downstairs brain springs into action, preparing to react quickly rather than fully considering alternatives in a more responsive, receptive state. His muscles might tense as he prepares to defend himself and, if necessary, attack with freeze and fight. Or he may run away in flight, or collapse in a fainting response. Each of these is a pathway of reactivity of the downstairs brain. And his thinking, rational self-control circuitry of the upstairs brain goes off-line, becoming unavailable in that moment. That's the key—we can't be in both a reactive downstairs state and a receptive upstairs state at the same time. The downstairs reactivity holds sway.

In this situation, you can appeal to your child's more sophisticated upstairs brain, and allow it to help rein in the more reactive downstairs brain. By demonstrating respect for your child, nurturing him with lots of empathy, and remaining open to collaborative and reflective discussions, you communicate "no threat," so the reptilian brain can relax its reactivity. In doing so, you activate the upstairs circuits, including the extremely important prefrontal cortex, which is responsible for calm decision making and controlling emotions and impulses. That's how we move from reactivity to receptivity. And that's what we want to teach our children to do.

So instead of fiercely demanding that your five-year-old calm down, you can help quiet and soothe the downstairs brain and instead bring the upstairs brain online by gently inviting him to be physically close to you and listening to whatever he's upset about. (If you're in a public place and your child is disturbing everyone around you, it may be necessary to take him outside while you attempt to appeal to his upstairs brain.)

Research supports this strategy of engaging the upstairs brain rather than enraging the downstairs. We've seen, for example, that when a person is shown a photo of a face that's angry or afraid, activity increases in a region of the downstairs brain called the amygdala (pronounced *uh-MIG-duh-luh*), which is responsible for quickly processing and expressing strong emotions, especially anger and fear. One of the amygdala's primary jobs is to remain alert and to sound an alarm anytime we are threatened, allowing us to act quickly. Interestingly, simply seeing a photograph of a person with an angry or frightened face causes the viewer's amygdala to activate. In fact, even if the viewer sees the photo so quickly that he or she isn't consciously aware of having seen the picture, a subliminal, instinctual, emotional reaction causes the amygdala to fire, or become active.

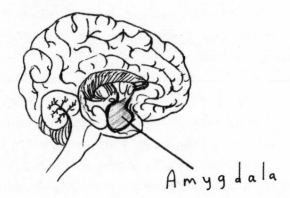

Amygdala

What's even more fascinating about the study is that when viewers were asked to label the emotion in the picture, and named it as fear or anger, their amygdala immediately became less active. Why? Because part of the upstairs brain—a part called the ventrolateral prefrontal cortex—took charge with the labeling and then processed

INSTEAD OF ENRAGING THE DOWNSTAIRS BRAIN...

ENGAGE THE UPSTAIRS BRAIN

the emotion, allowing the thinking, analytical part of the brain to take over and soothe the irritated lower regions, rather than letting the reactive, emotional downstairs brain dominate and dictate the person's feelings and responses. This is a classic example of the "name it to tame it" strategy we discuss in detail in *The Whole-Brain Child*. Simply by naming the emotion, a person feels her levels of fear and anger decrease. That's how the upstairs brain can calm the downstairs brain. And that's a skill that can last a lifetime.

This is what we want to do for our kids when they become upset and act out: help them engage their upstairs brain. The prefrontal part of the upstairs brain actually has soothing fibers that can calm the lower regions when they are reactive. The key is to grow them well in our children, and to activate them in a moment of distress by first connecting before redirecting. We want our kids to develop the internal skill to calm the storm and reflect on what's happening inside.

Think back about the functions of the upstairs brain: good decision making, control over emotions and body, flexibility, empathy, self-understanding, and morality. These are the aspects of our kids' character we want to develop, right? As we put it in *The Whole-Brain Child*, we want to engage the upstairs brain, rather than enraging the downstairs brain. Engage, don't enrage. When we enrage the downstairs brain, that's usually because *our* amygdala is firing as well. And guess what the amygdala wants to do. Win! So when the amygdalae in both the parent and the child are firing at top speed, both looking to win, it's virtually always going to be a dramatic battle that ends with both sides losing. No one will win, and relational casualties will litter the battlefield. All because we enraged the downstairs, rather than engaging the upstairs.

To use a different metaphor, it's as if you have a remote control for your child, and you have the power, at least to some extent, to determine what kind of a response you'll receive when you two interact. Press the engage button—the "calm down and think" button—and you'll appeal to the upstairs brain, activating a calming response. But push the enrage

BATTLE OF THE AMYGDALAE: NOBODY WINS

button—the "freak out and escalate emotions" button—by using threats and demands, and you'll be practically begging for the fighting part of the brain to click into action. You'll poke the lizard and get a reptilian, reactive response. It's up to you which button to press.

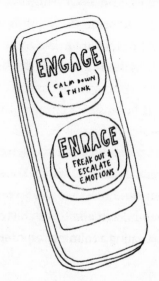

Remember, none of this is to excuse parents from the responsibility of setting boundaries and clearly communicating expectations. We'll give you lots of practical suggestions for doing so in the coming pages. But as you set those boundaries and communicate those expectations, you'll make things much easier on yourself, your child, and anyone else within shouting range if you appeal to your child's wiser and receptive self and her upstairs brain, as opposed to her lizard reactiveness and her downstairs brain.

What's even more exciting is what happens *after* we appeal to the upstairs brain. When it gets engaged repeatedly, it becomes strong. Neurons that fire together wire together. So when a child is in an upset state of mind and we invite the upstairs brain to become active, we create a functional linkage between that dysregulated state and an activation of the part of her brain that brings her back into a well-regulated state. We can likely grow those soothing fibers that extend from the prefrontal upstairs brain into the downstairs brain.

That means the more we appeal to our child's more integrated nature—the more we ask her to think before she acts or to consider someone else's feelings, the more we ask her to act ethically or empathically—then the more she'll use her upstairs brain, and the stronger it will become, because it is building connections and becoming more integrated with the downstairs areas. Using her upstairs brain will more and more become her accessible pathway, her automatic default, even when emotions run high. As a result, she'll become better and better at making good decisions, handling her emotions, and caring for others.

Applying the Brain C's

Let's talk now about what the three Brain C's—changing, changeable, and complex—look like in action. When Nina freaked out on the porch step, Liz's first instinct was to logically explain how the transportation decisions had been made: "Your sister's school is right by my work." She could have gone on to explain that Tim had more time to drive Nina to her school, and that Nina had just yesterday been asking for more time with her father. All of these statements were true, and rational.

However, as we've said, when a child is in the throes of a meltdown, logic will often be ineffective, sometimes even counterproductive. This is what Liz recognized as she looked at her fury-filled daughter. In effect, what she realized was the first of the three Brain C's: Nina's brain was *changing*. It was developing. Not developed, *developing*. Which meant that Liz needed to be patient with her little girl and not expect her to consistently control herself like an adult, or even like an older child. She took a deep breath and worked to remain calm, despite the stress being produced by the unreasonable four-year-old, the impatient seven-year-old, and the ever-ticking clock.

Just as important in this situation was the second Brain C, that the brain is *changeable*. Liz understood that the way she and her husband handled each situation with their daughters wired the girls' developing brains, for good or for bad. So in this moment of awareness, Liz resisted the urge she currently felt, which was to hurriedly and even aggressively pick up her crying daughter, march her to Tim's car stomping all the way, strap her into her car seat, and slam the door.

By the way, if you recognize yourself in the anger-filled depiction of how Liz wanted to handle the situation, you're not alone. We've all been there. (See "When a Parenting Expert Loses It" at the back of the book.) Caring parents will often condemn themselves over every little mistake they make, or for every time they miss an opportunity to approach a difficult moment from a Whole-Brain perspective. We urge you to listen to this internal critic only long enough to gain some awareness so that you can do better the next time, but then be generous and forgiving of yourself. Of course you want to do your best with and for your kids. But as we'll explain in detail in the book's conclusion, even parental mistakes can be extremely valuable for our kids—we can teach them we are all human, and we can take responsibility for what happens and make a repair. That's an essential teaching experience for all children.

Liz was human and a parent, so of course she made her share of mistakes, as we all do. But in this instance she disciplined from a No-Drama, Whole-Brain frame of mind and made an intentional decision to take a moment and be there emotionally for her young daughter. By this point the family was less than one minute behind schedule. And Liz realized that even though Nina's feelings seemed dramatic, they were real. She needed her mom right then. So Liz denied the impulse to do what was easiest and quickest, and again pulled her daughter close to her.

As for specifically how she responded to the situation, that's where the third Brain C—*complex*—comes in. Liz understood her daughter well enough to know better than to enrage the downstairs brain. It was plenty active already. Instead, she needed to engage Nina's upstairs brain. The first step, though, had to be to connect. Before redirecting, we always connect. That's what Liz was doing when she held her daughter. Yes, she was in a hurry, but nothing positive could happen until Nina calmed down some, which didn't take long once she was in her mother's arms. In just a few seconds Liz felt Nina take a deep breath and her little body begin to soften.

If Nina were your child, you might have handled this situation in one of a few ways, depending on your style and her temperament. Like Liz, you would probably seek as your first goal to help your daughter calm down, so that her upstairs brain would come back online and she could listen to reason. You might promise to get up early tomorrow morning so you'd have time to take her to school. Or you might assure her that you'd ask your boss if you could leave work early this afternoon so you could pick up your daughter and then have some one-on-one special time with her. Or you might offer to tell her a story on speakerphone from your car as her dad drove her to school.

As it turned out, Liz tried several of these strategies, all to no avail. No creative inspiration hit the mark. Nina was having none of it.

Aren't you glad we didn't use an example in which that situation worked out nicely and perfectly? You're relieved, aren't you, because you know it doesn't always go that way. No matter how skillfully we handle a situation, and no matter how cognizant we remain of important information like the three Brain C's, at times our kids still don't do things the way we'd like. They don't pick up their toys. They don't automatically apologize to their brother. They don't calm down. Which is exactly what happened here. Nina would not cooperate. Listening to her feelings, holding her, coming up with a plan . . . nothing did the trick.

But Liz still had to leave for work, and the kids had to get to school. So, remaining calm and empathic—that's our goal—she explained that they had to go, and that Tim would drive her to school this morning as planned: "I know you're sad, and I understand that you want to ride with me. I would like that, too. But we can't make that work today. Would you like to climb in or would you like Daddy to help you get in the car now? Daddy will be with you to comfort you on the way to school. I love you and I'll see you this afternoon." And with that, the front-porch situation ended, with Tim holding a crying Nina as he carried her to his car.

Notice what we're acknowledging here. No-Drama Discipline can't ensure that your kids will act the way you'd like every time you address their behavior. The Whole-Brain approach definitely gives you a much better chance of achieving the short-term goal of encouraging cooperation from your children. It also helps remove or at least reduce the most explosive emotions in the situation, deescalating the drama and thus avoiding the harm and hurt that result when a parent yells or personalizes the issue. But it won't *always* be effective at getting the exact behavior you hope for. Kids are human beings, after all, who have their own emotions, desires, and agendas; they're not computers we program to do what we want. But at the very least, as we're sure you'll agree after you read the following chapters, No-Drama Discipline gives you a much better chance of communicating with your children in ways that feel better to both of you, build trust and respect between you, and decrease the drama in most discipline situations.

What's more, a Whole-Brain approach provides a way to show

our kids how much we love and respect them, even as we discipline them. They know—and we reinforce it over and over throughout their lives—that when they're upset or acting inappropriately, we're going to be there for them. And with them. We don't turn our back or reject them when they're distressed. We don't say, or even imply, that their happiness is a condition they must meet to receive our love. *No-Drama Discipline allows us to communicate to our children, "I'm with you. I've got your back. Even when you're at your worst and I don't like the way you're acting, I love you, and I'm here for you. I understand you're having a hard time, and I am here."* No parent can communicate this message all the time in every scenario. But we can send it consistently and repeatedly, so there's never any question in our children's minds.

This kind of predictable, sensitive, loving, relational discipline allows kids to feel safe. As a result, they have the freedom to become independent individuals whose brains are wired in such a way that they are better able to think through decisions, comprehend what they actually feel about a situation, consider others' perspectives, and come to a sound conclusion on their own. In other words, the experiences of emotional and physical safety give them the capacity to act responsibly and make good choices. In contrast, a parenting style focused on control and fear, stressing that a child needs to toe the line all the time, undermines that feeling of safety. If a child lives in constant worry that he might mess up and make his parents unhappy or that he'll be punished, he won't feel the freedom to do all the things that grow and strengthen his upstairs brain: considering others' feelings, exploring alternative actions, understanding himself, and trying to make the best decision in a given situation. We don't want our discipline to cause our children to focus all of their energy and neural resources on making us happy or staying out of trouble. Instead, we want our discipline to help grow our kids' upstairs brains. And that's just what No-Drama Discipline does.

No-Drama Discipline Builds the Brain

The three Brain C's lead to one crucial and undeniable conclusion, which is the central notion of this chapter: No-Drama Discipline actually helps build the brain. That's right. It's not *only* that a Whole-Brain approach can defuse difficult and highly charged situations with your kids. Or that it will help you build your relationship with them as you more clearly communicate how much you love them and that they are safe, even as you set boundaries for their behavior. All that's true; the discipline principles and strategies we'll show you in the coming pages really do offer all of those benefits, making your day-to-day life easier and less stressful while nurturing your relationship with your child.

But beyond all that, No-Drama Discipline actually builds a child's brain. It strengthens neural connections between the upstairs and downstairs parts of the brain, and these connections lead to personal insight, responsibility, flexible decision making, empathy, and morality. The reason is that when we help strengthen the connective fibers between the upstairs and downstairs, the higher parts of the brain can communicate with and override a child's primitive impulses more and more often. And our disciplinary decisions go a long way toward determining how strong those connections are. *The way we interact with our kids when they're upset significantly affects how their brains develop, and therefore what kind of people they are, both today and in the years to come.* This is how our way of communicating with our children impacts their internal skills, which are embedded in the connections in their changing, changeable, and complex brains!

It makes perfect sense when you think about it. Every time we give a child the experience of exercising his upstairs brain, it gets stronger and more fully developed. When we ask him questions that develop insight into himself, he becomes more insightful.

When we encourage her to empathize with someone else, she becomes more empathic.

When we give a child the opportunity to *decide* how he should act, rather than simply telling him what he should do, he becomes a better decision maker.

And that's one of the ultimate goals of parenting, isn't it? That our kids become more insightful, empathic, and able to make good decisions *on their own*? You know the old saying: "Give a man a fish, and he'll eat for a day. Teach a man to fish, and he'll eat for a lifetime." Our ultimate goal isn't that our children do what we want them to do because we're watching them or telling them what to do. (That would be fairly impractical, after all, unless we plan on living and going to work with them for the rest of their lives.) Rather, we want to help them learn to make positive and productive choices on their own in whatever situation they face. And that means we need to view the times they misbehave as opportunities to give them practice building important skills and having those experiences wired into the brain.

Building the Brain by Setting Limits

This perspective can completely change the way we look at the opportunities we have to help our kids make better choices. When we set limits, we help develop the parts of the upstairs brain that allow children to control themselves and regulate their behaviors and their body.

One way to think about it is that we're helping our kids develop the ability to shift between the different aspects of what's called the autonomic nervous system. One part of the autonomic nervous system is the sympathetic branch, which you can think of as the "accelerator" of the system. Like a gas pedal, it causes us to react with gusto to impulses and situations, as it primes the body for action. The other part is the parasympathetic branch, which serves as the "brakes" of the system and allows us to stop and regulate ourselves and our impulses. Keeping the accelerator and the brakes in balance is key for emotional regulation, so when we help children develop the capacity to control themselves even when they're upset, we're helping them learn to balance these two branches of the autonomic nervous system.

Purely in terms of brain functioning, sometimes an activated accelerator (which might result in a child's inappropriate and impulsive action) followed by the sudden application of brakes (in the form of parental limit setting) leads to a nervous system response that may cause the child to stop and feel a sense of shame. When this happens, the physiologic manifestation might result in avoiding eye contact, feeling a heaviness in her chest, and possibly experiencing a sinking feeling in her stomach. Parents might describe this by saying she "feels bad about what she's done."

This initial awareness of having crossed a line is extremely healthy, and it's evidence of a child's developing upstairs brain. Some scientists suggest that limit setting that creates a "healthy sense of shame" leads to an internal compass to guide future behavior. It means she's beginning to acquire a conscience, or an inner voice, along with an understanding of morality and self-control. Over time, as her parents repeatedly help her recognize the moments when she needs to put on the brakes, her behavior begins to change. It's more than simply learning that a particular action is bad, or that her parents don't like what she's done, so she'd better avoid that action or she'll get in trouble. More occurs within this child than just learning the rules of good vs. bad or acceptable vs. unacceptable.

Rather, her brain actually changes, and her nervous system gets wired to tell her what "feels right," which modifies her future behavior. New experiences wire new connections among her neurons, and the changes in the circuitry of her brain fundamentally and positively alter the way she interacts with her world. The way her parents help this process along is by lovingly and empathically teaching her which behaviors are acceptable and which aren't. That's why it's essential that we set limits and that our children internalize "no" when necessary, particularly in the early years, when the regulatory circuits of the brain are wiring up. By helping them understand the rules and limits in their respective environments, we help build their conscience.

This is often difficult for a loving parent. We want our kids to be happy, and we like it when they receive what they desire. Plus we're aware of how quickly a pleasant situation can devolve when a child doesn't get what he wants. However, if we truly love our kids and want what's best for them, we need to be able to tolerate the tension and discomfort they (and we) may experience when we set a limit. We want to say yes to our children as often as possible, but sometimes saying no is the most loving thing we can do.

One caveat here: Many parents say no, or a form of it, far too often. They say it automatically, often when it's not necessary. *Stop touching that balloon. No running. Don't spill.* Our point here isn't that we want our kids to hear the word "no" a lot. In fact, much more effective than an outright no is a yes with a condition: "Yes, you can take a bath later" or "Yes, we'll read another story, but we'll need to do it tomorrow." *The point, in other words, is not to make a point of saying no, but to understand the importance of helping kids recognize limits so that they become increasingly better at putting on the brakes themselves when necessary.*

A second caveat is important to note here, too. When limit-setting and "no" are accompanied by parental anger or negative comments that assault a child, the "healthy, developmental shame" of a child simply learning to curb his or her behavior now is transformed into more complicated "toxic shame" and humiliation. One view proposes that toxic shame involves not simply the sense of having done something wrong, which can and needs to be corrected, but the painful sense that one's inner self is defective. And this belief that the self is damaged is felt to be an unchangeable condition of the child—not a behavior that can be modified. Some researchers consider this move from "behavior to be changed in the future" to a "self that is fundamentally flawed" as the outcome for children who experience repeated parental hostility in response to their behavior. Toxic shame and humiliation can continue through childhood and into adulthood, even beneath the surface of awareness, leaving individuals with a hidden "secret" that they are permanently and deeply defective. A cascade of negative consequences—having trouble with close relationships that might reveal this hidden secret, feeling unworthy, being driven to succeed in life but never feeling satisfied—can then dominate the individual's life. You as a parent can avoid giving your child this negative cascade of toxic shame by learning how to create needed structure without humiliating your child. That's an achievable goal, and we are committed to making that path available to you if you choose it.

What it all comes down to is that No-Drama Discipline encourages kids to look inside themselves, consider the feelings of others, and make decisions that are often difficult, even when they have the impulse or desire to do things another way. It allows children to put into practice the emotional and social abilities we want them to understand and master. It allows you to create structure with respect. When we're willing to lovingly set a boundary—just like when we discipline with an awareness that our children's brains are changing,

INSTEAD OF AN OUTRIGHT NO...

TRY A YES WITH A CONDITION

changeable, and complex—we help create neural connections that improve our kids' capacity for relationships, self-control, empathy, personal insight, morality, and much, much more. And they can feel good about who they are as individuals while learning to modify their behavior.

All of this leads to an exciting conclusion for parents: every time our children misbehave, they give us an opportunity to understand them better, and get a better sense of what they need help learning. Children often act out because they haven't yet developed skills in a particular area. So when your three-year-old pulls her classmate's hair because he got the first Dixie cup full of fish crackers, she's actually telling you, "I need to build skills in waiting my turn." Likewise, when your seven-year-old becomes defiant and calls you "Fart-face Jones" after you tell him it's time to leave his playdate, he's actually saying, "I need skill building when it comes to handling myself well and communicating my disappointment respectfully when I don't get my way." By misbehaving, kids actually communicate to us what

WHAT A PARENT SEES:

WHAT THE CHILD IS REALLY SAYING:

they need to be working on—what has not yet been developed or what specific skills they need practice with.

The bad news is that it's rarely much fun, either for the child or for the parent. The good news is that we get information we might not otherwise receive. The even better news is that we can then take intentional steps to give our kids experiences that help them improve on their ability to share, think of others, speak kindly, and so on. We're not saying that when your children don't handle things well, you should necessarily celebrate. ("Yay! An opportunity to help a brain develop optimally with my intentional response!") You're probably not going to enjoy discipline, or look forward to future meltdowns. But when you realize that these "misbehavior moments" aren't just miserable experiences to endure, but actually opportunities for knowledge and growth, you can reframe the whole experience and recognize it as a chance to build the brain and create something meaningful and significant in your child's life.

CHAPTER 3

From Tantrum to Tranquility: Connection Is the Key

Michael heard voices rising in his sons' room but was watching the basketball game on TV and decided to wait for a commercial before investigating. Big mistake.

His eight-year-old, Graham, and Graham's friend James had spent the last thirty minutes carefully organizing and categorizing Graham's hundreds of Lego pieces. Graham had used his allowance to buy a fishing tackle box, and he had designated a different compartment for every Lego head, torso, helmet, sword, light saber, wand, axe, and anything else the creative geniuses from Denmark could dream up. The boys were in organizational heaven.

The problem was that Michael's five-year-old, Matthias, had been feeling increasingly left out by Graham and James. The three boys had begun the project together, but the older boys eventually felt that Matthias didn't quite understand their complex categorical system. As a result, they weren't allowing him to participate in the activity.

Cue the rising voices.

Michael never made it to the commercial. The shouting let him know that he needed to intervene immediately, but he wasn't quick enough. When he was still three steps away from the boys' room—

three short steps!—he heard the unmistakable sound of hundreds of plastic Lego pieces exploding across a hardwood floor.

Three steps later he witnessed the mayhem and carnage. It was a complete massacre. Decapitated heads littered the entire room, lying next to armless bodies and weapons both medieval and futuristic. A rainbow of chaos stretched from the doorway to the closet on the other side of the room.

Next to the upended tackle box stood Michael's huffing, red-faced five-year-old, looking at him with eyes that were somehow both defiant and terrified. Michael turned to his older son, who yelled, "He ruins *everything*!" and ran from the room in tears, followed by a sheepish-looking and uncomfortable James.

Talk about a discipline moment. Both of his boys were now bawling, a friend was caught in the crossfire, and Michael himself felt furious. Not only had Matthias destroyed all the work the older boys had done, but now there was a huge mess to clean up in the room. (If you've ever felt the pain of stepping on a Lego piece, you know why it wasn't an option to leave the bits spread out on the floor.) And he was missing the game.

Michael decided he'd go check on the older boys in a minute and address Matthias first. His initial inclination was to stand over his young son, wag his finger in his son's face, and scold him for dumping the tackle box. In his anger he wanted to offer immediate consequences. He wanted to shout, "Why did you *do* this?" He wanted to say something about never again getting to participate in Graham's playdates, then add, "Do you see why they didn't want you to play with their Legos?"

Luckily, though, the thinking part of Michael (his upstairs brain) took over, and he addressed the situation from a Whole-Brain perspective. What triggered the more mature and empathic approach was his recognition of how much his little boy *needed* him right then. Of course Michael would have to address Matthias's behavior. And yes, he'd obviously need to be a bit more proactive next time in attending to the situation before it spun out of control. He'd want to help Matthias think about how Graham felt, and understand that our actions often impact other people in significant ways. All of this teaching, all of this redirection, was absolutely necessary.

But not right now.

Right now, he needed to connect.

Matthias was completely dysregulated emotionally, and he needed his dad to soothe the hurt feelings, sadness, and anger that came from being criticized for being too little to understand and from being excluded. This was not the time to redirect, to teach, or to talk about family rules and respect for others' property. It was time to connect.

So Michael knelt down and opened his arms, and Matthias fell into them. Michael held him as he sobbed, rubbed his back, and said nothing other than an occasional "I know, buddy. I know."

A minute later Matthias looked up at him, his eyes shiny with tears, and said, "I spilt the Legos."

In response, Michael laughed a little and said, "I'd say you did more than that, little man!"

Matthias cracked a small smile, and at that point Michael knew he could now proceed to the redirecting part of the discipline and help Matthias understand some important lessons about empathy and appropriate expressions of big feelings. He was now *capable* of hearing his father. Michael's connection and comfort had allowed his son to move out of a reactive state and into a receptive one, where he could hear his dad and really learn.

Notice that connecting first is not only more relational and loving. Yes, it allows parents to attune to their children, as Michael did here, and be emotionally responsive when they're upset and dysregulated. That enables the child to "feel felt," which is the inner sense of being seen and understood that transforms chaos into calm, isolation into connection. Connecting first is a fundamentally loving way to discipline. But notice how much more *effective* a No-Drama disciplinary approach can be as well. It's not that a lecture would have been *wrong* as Michael's initial response to the situation. Our point here isn't about the rightness or wrongness of parenting approaches (although we'd definitely argue that a Whole-Brain approach is fundamentally more loving and compassionate). The point is that Michael's connect-first tactic achieved the two goals of discipline—gaining cooperation and brain building—extremely effectively. It allowed learning to occur, teaching to be effective, and connection to be established and maintained. His approach let Michael get his son's attention, and to do so quickly and without drama, so they could talk about Matthias's behavior *in such a way that he could listen*. Plus, it could help build Matthias's brain, because he could now hear Michael's points and understand the important lessons his father was teaching him. In addition, Michael modeled for his son attuned connection and showed him that there are calmer, more loving ways to interact when you're upset with someone. And all of this happened because Michael connected first, before redirecting.

Proactive Parenting

We'll talk in just a minute about why connection is such a powerful tool when our kids are upset or having trouble making good decisions. Michael obviously used it effectively. But by being just a bit slow to respond to the situation—three short steps!—he missed an opportunity to avoid the entire disciplinary process completely.

It really is true. At times we can avoid having to discipline at all, simply by parenting *proactively,* rather than *reactively.* When we parent proactively, we watch for times we can tell that misbehavior and/or a meltdown is in our children's near future—it's just over the horizon of where they are right now—and we step in and try to guide them around that potential landmine. Michael wanted to make it to the next commercial, so he didn't respond quickly enough to the signs that trouble was beginning to surface in his sons' room.

Parenting proactively can make all the difference. When, for example, your sweet and usually compliant eight-year-old is getting ready to go to her swim lesson, you might notice that she overreacts a bit when it's time to apply sunscreen: "Why do I have to use sunscreen every day?" Then while you're getting her little brother ready, she sits down at the piano for a minute to play one of her songs. But she misses a couple of notes, then slams her fist on the keyboard in frustration.

You could interpret these actions as isolated incidents and overlook them. Or you could recognize them for the warning flags they probably are. You might remember that this particular daughter gets especially upset when she's hungry, so you might stop what you're doing and set an apple in front of her. When she looks up at you, you can offer her a knowing smile as a reminder of this tendency of hers, and hopefully she'll nod, eat the apple, and move back into a place of self-control.

Granted, sometimes no obvious signs present themselves before our kids make bad decisions and act in ways that aren't ideal. But other times we can read our children's cues and take proactive steps to stay ahead of the discipline curve. That might mean giving a warning five minutes before having to leave the park, or enforcing a consistent bedtime so your kids don't get too tired and grumpy. It might mean starting to tell a preschooler a suspenseful story and then pausing it, explaining that you'll tell what happens next once she's in her car seat. Or maybe it means you step in to begin a new game when you hear that your children are moving toward significant conflict with each other. It might mean telling a toddler, with a voice full of intriguing energy, "Hey, before you throw that french fry across the restaurant, I want to show you what I have in my purse."

INSTEAD OF PARENTING REACTIVELY...

PARENT PROACTIVELY

Another way to parent proactively is to HALT before responding to your kids. When you see your child's behavior trending in a direction you don't like, ask yourself, "Is he **h**ungry, **a**ngry, **l**onely, or **t**ired?" It may be that all you need to do is to set out some raisins, listen to his feelings, play a game with him, or help him get more sleep. Sometimes, in other words, all you need is a bit of forethought and planning ahead.

IS YOUR CHILD TOO:

HUNGRY?

ANGRY?

LONELY?

TIRED?

Parenting proactively isn't easy, and it takes a fair amount of awareness on your part. But *the more you can watch for the beginnings of negative behaviors and head them off at the pass, the less you'll end up having to pick up the literal or figurative pieces, meaning you and your children will have more time simply to enjoy each other.*

As we all know, though, sometimes misbehavior just happens. Oh, does it happen. And no amount of proactivity can prevent it. That's when it's time to connect. We have to fight the urge to immediately punish, lecture, lay down the law, or even positively redirect right away. Instead, we need to *connect*.

Why Connect First?

Let's get more specific and talk about *why* connection is so powerful. We'll look at three primary benefits—one short-term, one long-term, and one relational—of making connection our first response when our kids have trouble controlling themselves and making good decisions.

Benefit #1: Connection Moves a Child from Reactivity to Receptivity

However we decide to specifically respond when our children misbehave, there's one thing we have to do: we must remain emotionally connected with them, even when—and perhaps especially when—we discipline. After all, *it's when our kids are most upset that they need us the most*. Think about it: they don't *want* to feel frustrated, enraged, or out of control. That's not only unpleasant, it's extremely stressful. Usually misbehavior is the result of a child having a hard time dealing with what's going on around her—and inside her. She's got all these big feelings she doesn't yet have the capacity to manage, and the misbehavior is simply the result. Her actions—especially when she's out of control—are a message that she needs help. They are a bid for assistance, and for connection.

So when children feel furious, dejected, ashamed, embarrassed, overwhelmed, or out of control in any other way, that's when we need to be there for them. Through connection, we can soothe their internal storm, help them calm down, and assist them in making better decisions. When they feel our love and acceptance, when they "feel felt" by us, even when they know we don't like their actions (or they

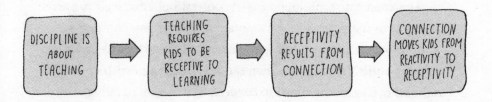

don't like ours), they can begin to regain control and allow their up-stairs brains to engage again. When that happens, effective discipline can actually take place. Connection, in other words, moves them out of a reactive state and into a state where they can be more receptive to the lesson we want to teach and to the healthy interactions we want to share with them.

So there's a great question we can ask ourselves before we begin redirecting and explicitly teaching: *Is my child ready? Ready to hear me, ready to learn, ready to understand?* If a child isn't ready, then more connection is most likely in order.

As we saw with Michael and his five-year-old, connection calms the nervous system, soothing children's reactivity in the moment and moving them toward a place where they can hear us, learn, and even make their own Whole-Brain decisions. When the emotional gauge gets turned up, connection is the modulator that keeps the feelings from getting too high. Without connection, emotions can continue to spiral out of control.

CONNECTION CALMS

Imagine the last time you felt really sad or angry or upset. How would it have felt if someone you love told you, "You need to calm down," or "It's not *that* big a deal"? Or what if you were told to "go be by yourself until you're calm and ready to be nice and happy"? These responses would feel awful, wouldn't they? Yet these are the kinds of things we tell our kids all the time. When we do, we actually *increase* their internal distress, leading to more acting out, not less. These re-sponses accomplish the opposite of connection, effectively *amplify-ing* negative states.

Connection, on the other hand, calms, allowing children to begin to regain control of their emotions and bodies. It allows them to "feel felt," and this empathy soothes the sense of isolation or being misun-derstood that arises with the reactivity of their downstairs brain and the whole nervous system: heart pounding, lungs rapidly breathing, muscles tightening, and intestines churning. Those reactive states are uncomfortable, and they can become intensified with further demands and disconnection. With connection, however, kids can make more thoughtful choices and handle themselves better.

What connection does, essentially, is to *integrate* the brain. Here's how it works. The brain, as we've said, is complex. (That's the third Brain C.) It's made up of many parts, all of which have different jobs to do. The upstairs brain, the downstairs brain. The left side and the right side. There are memory centers and pain regions. Along with all the systems and circuitry of the brain, these parts of our brain have their own responsibilities, their own jobs to do. When they work together as a coordinated whole, the brain becomes integrated. Its many parts can perform as a team, accomplishing more and being more effective than they could working on their own.

As we explained in *The Whole-Brain Child*, a good image to help understand integration is a river of well-being. Imagine you're in a canoe, floating along in a peaceful, idyllic river. You feel calm, relaxed, and ready to deal with whatever comes along. It's not necessarily that everything's perfect or going your way. It's more that you're in an integrated state of mind—you're calm, receptive, and balanced, and your body feels energetic and at ease. Even when things don't work out the way you'd like, you can flexibly adapt. That's the river of well-being.

Sometimes, though, you're not able to stay in the flow of the river. You veer too far to one bank or the other. One side of the river represents chaos. Near this bank are dangerous rapids that make life feel frenzied and unmanageable. When you're near the chaos bank, you're easily upset, and even minor obstacles can leave you spinning out of control. You might experience overwhelming emotions such as high anxiety or intense anger, and you might notice that your body feels chaotic, too, with tense muscles, a rapid heartbeat, and a furrowed brow.

The other bank is no less unpleasant, because it represents rigidity. Here you get stuck desiring or expecting the world to operate in one particular way, and you're unwilling or unable to adapt when it doesn't. In your effort to impose your own vision and desires on the world around you, you find that you won't, or possibly even can't, compromise or negotiate in any meaningful way.

So chaos is on one bank, rigidity on the other. The two extremes offer either a lack of control or so much control that there's no flexibility or adaptability. And both extremes keep you out of the peaceful flow of the river of well-being. Whether you're chaotic or rigid, you're missing out on the opportunity to enjoy mental and emotional health, to feel at ease with the world.

Think about the river of well-being in relation to your kids. Almost always, when children act up or feel upset, they will display evidence of chaos, rigidity, or both. When a nine-year-old freaks out

about an oral presentation at school the next day and ends up ripping up her notes as she sobs that she'll never be able to memorize her opening, she's succumbed to chaos. She's crashed into the bank, far from the smooth-flowing river of well-being. Similarly, when a five-year-old stubbornly insists on another bedtime story or refuses to get in the tub until he finds his most special wristband, he's right up against the rigidity bank. And remember Nina from the last chapter? When she fell apart because her mom told her that her dad would be driving her to school that morning, then refused to consider any alternative perspectives on the situation, she was zigzagging back and forth between chaos and rigidity, never getting to enjoy the peaceful flow in the center of the river of well-being.

So that's what connection does. It moves children away from the banks and back into the flow, where they experience an internal sense of balance and feel happier and more stable. Then they can hear what we need to tell them, and they can make better decisions. When we connect with a child who feels overwhelmed and chaotic, we help move her away from that bank and into the center of the river, where she can feel more balanced and in control. When we connect with a child who's stuck in a rigid frame of mind, unable to consider alternative perspectives, we help him integrate so that he can loosen his unyielding grip on a situation and become more flexible and adaptive. In both cases, connection creates an integrated state of mind, and the opportunity for learning.

We'll get much more specific in the next chapter about practical ways to connect with your children when they're upset. The basic approach, though, usually entails listening and providing lots of verbal and nonverbal empathy. This is how we attune to our children, tuning in to the inner life of their mind—to their feelings and thoughts, to their perceptions and memories, to what has inner subjective meaning in their lives. This is tuning in to *the mind beneath their behavior*. For example, one of the most powerful ways we connect with our children is simply by physically touching them. A loving touch—as simple as a hand on an arm or a rub on the back or a warm embrace—releases feel-good hormones (like natural oxytocin and opioids) into our brain and body, and decreases the level of our stress hormone (cortisol). When your children are feeling upset, a loving touch can calm things down and help you connect, even during moments of high stress. This is connecting with their inner distress, not simply reacting to their outwardly visible behavior.

CONNECTION MOVES A CHILD FROM REACTIVITY TO RECEPTIVITY

Notice that this was the first thing Michael did when he looked at his young son in the middle of the Lego carnage: he sat down and held him.

In doing so, he began to pull Matthias's tiny canoe away from the bank of chaos and back into the peaceful flow of the river. Then he listened. Matthias didn't need to say much: "I spilt the Legos." With that he could begin to move on. Sometimes children will need to talk much more, and to be listened to for much longer. Or sometimes they don't want to talk. And sometimes it can be as quick as it was here. Nonverbal touch, an empathic statement—"I know, buddy"—

and a willingness to listen. That's what Matthias needed in order to return some equilibrium to his young brain and impulsive body. Once that happened, his father could begin to teach him by talking about the lessons at hand.

Even though Michael wasn't thinking in these terms, what he was doing was using his relationship, his connecting communication, to help bring integration to Matthias's brain, so that his upstairs brain and his downstairs brain could work together, and so that the right and left sides of his brain could work together. When Matthias became furious with the older boys, his downstairs brain completely took over, disabling his upstairs brain. The instinctive, reactive lower parts of his brain became so active that he lost access to the higher parts of the brain, the ones that help him think about consequences and consider others' feelings. These two parts of his brain were not working together. In other words, his brain in that moment was dis-integrated, and the result was the Lego massacre. By offering a nonverbal gesture instead of just a bunch of logical, left-brained words, Michael was able to connect with Matthias's right brain, the side more directly connected to and also flooded by the downstairs brain. Right and left, downstairs and upstairs, Matthias's brain was ready to become more coordinated and balanced in its movement toward integration. Connection integrated his emotion-focused downstairs brain and his thinking-oriented upstairs brain and allowed Michael to achieve the short-term goal of gaining cooperation from his son.

Benefit #2: Connection Builds the Brain

As we explained in the previous chapter, No-Drama Discipline builds the brain of a child by improving his capacity for relationships, self-control, empathy, personal insight, and much more. We discussed the importance of setting limits, creating structure, and helping children build internal controls and impulse inhibition by internalizing "no." This is how we use our relationship with our children to build their brains' executive functions. We also discussed other ways to develop a child's relational and decision-making abilities. Each inter-action with our kids offers the opportunity to build their brains and further their capacity to be the kind of people we hope they'll be.

And it all begins with connection. In addition to the short-term benefit of moving them from reactivity to receptivity, connecting during a disciplinary interaction also impacts children's brains in ways that will have long-term effects as they grow up. When we offer comfort when our kids are upset, when we listen to their feelings, when we communicate how much we love them even when they've messed up: when we respond in these ways, we significantly impact the way their brains develop and the kind of people they'll be, both now and as they move into adolescence and adulthood.

In upcoming chapters we'll talk more about redirection, including the explicit lessons we teach and the behaviors we model as we interact with our children. Obviously, a child's brain will be greatly impacted by what we communicate to him when we respond to misbehavior. And it will also be changed by what we model with our own actions in the moment. Whether consciously or subconsciously, a child's brain will assimilate all kinds of information based on the parental response to any situation. The more pertinent point here is about connection, and how parents change and even build children's brains based on what children experience in that disciplinary moment.

To put it in more neurological terms, connection strengthens the connective fibers between the upstairs and downstairs brain so that the higher parts of the brain can more effectively communicate with and override the lower, more primitive impulses. We nickname these fibers connecting upper and lower brain areas the "staircase of the brain." The staircase integrates upstairs and downstairs and benefits the region of the brain called the prefrontal cortex. This key area of the brain helps create the executive functions of self-regulation, in-cluding balancing our emotions, focusing our attention, controlling our impulses, and connecting us empathically with others. As the prefrontal cortex develops, children will be better able to put into practice the social and emotional skills we want them to develop and ultimately to master as they move through our home and out into the larger world.

To put it simply, integration in a relationship creates integration in the brain. An integrated relationship develops when we honor differences between ourselves and others, and then connect through compassionate communication. We empathize with another person, feeling their feelings and understanding their point of view. In this connection, we respect another person's inner mental life but do not become the other person. This is how we remain differentiated individuals but also connect. Such integration creates harmony in a relationship. Amazingly, interpersonal integration can also be seen at the heart of how parent-child relationships cultivate integration in the child's brain. This is how differentiated regions—like left and right, or up and down—remain unique and specialized but also become linked. Regulation in the brain depends upon the coordination and balance of various regions that emerge from integration. And such neural integration is the basis for executive functions, the capacity to regulate attention, emotions, thoughts, and behavior. That's the secret of the sauce! Interpersonal integration cultivates internal neural integration!

So that's the long-term benefit of connection: through relationships, it creates neural linkages and grows integrative fibers that literally change the brain and leave our kids more skilled at making good decisions, participating in relationships, and interacting successfully with their world.

Benefit #3: Connection Deepens the Relationship with Your Child

So connection offers the short-term benefit of moving kids from reactivity to receptivity, and the long-term benefit of building the brain. The third benefit we want to highlight is a relational one: connection deepens the bond between you and your child.

Moments of conflict can be the most difficult and precarious times in any relationship. They can also be among the most important. *Of course* our kids know we're there for them when we're snuggling and reading a book together, or when we show up and cheer at their performances. But what about when tension and conflict arise? When we have incompatible desires or opinions? These moments are the real test. How we respond to our children when we're not happy with their choices—with loving guidance? with irritation and criticism? with fury and a shaming outburst?—will impact the development of our relationship with them, and even their own sense of self.

It's not always easy to even *want* to connect when our kids misbehave, or when they're acting their ugliest and most out of control. Connecting might be the last thing in the world you want to do when a fight breaks out between your kids on a quiet airplane, or when they whine and complain about not getting a better treat after you've just taken them to the movies.

But connection should be our first response in virtually any disciplinary situation. Not only because it can help us deal with the problem in the short term. Not only because it will make our children better people in the long term. But also, and most important, because it helps us communicate how much we value the relationship. We know that our children have changing, changeable, and complex brains, and that they need us when they're struggling. The more we respond with empathy, support, and listening, the better it will be for our relationship with them.

Tina recently attended a birthday party with her six-year-old at his friend Sabrina's house. Her parents, Bassil and Kimberly, walked the guests out at the end of the party. When they returned to the living room of the house, they were met with a surprise. Here's how Kimberly put it in an email to Tina:

> After the party, Sabrina went into the house and opened all of her gifts *unsupervised*. So I couldn't write down who gave her what. It was pandemonium! I managed to piece together most of the items because my daughter Sierra had been there when she opened them. Before Sabrina writes out the thank-you

cards, I'd like to get this clarified. Did JP get her the kaleido-scope chalk? I'm sure Miss Manners would disapprove of my tactics, but I'd rather get it right than be nonspecific!

In this situation, we could certainly empathize with a tired parent for not handling herself well when she returned to the living room to find recently opened toys everywhere and torn wrapping paper littering the entire floor. After all, Kimberly had just hosted a fun but loud, entertaining but chaotic birthday party for fifteen six-year-olds and their parents and siblings. The circumstances were ripe for a parental meltdown, highlighted by lots of yelling about a spoiled kid who couldn't even wait until the party was over before ripping into the presents like a wild animal tearing into meat.

By maintaining her own self-control, though, Kimberly was able to address the situation from a No-Drama, Whole-Brain frame of mind, which led her to begin with—you guessed it—connection. Rather than launching into a lecture or a tirade, she connected with her daughter. She first acknowledged how fun it was to have had the

CONNECTION DEEPENS YOUR RELATIONSHIP WITH YOUR CHILD

party, and now to get to open all of the presents. She even sat patiently as Sabrina showed her the set of fake moustaches she was so excited about. (You'd have to know Sabrina.) And then, once Kimberly had connected, she spoke with her daughter, teaching her what she wanted her to know about presents and waiting and thank-you notes. That's how connection created an integrative opportunity, building a stronger brain and strengthening a relationship.

Will you be able to connect first every time your kids mess up or lose control of themselves? Of course not. We certainly don't with our own kids. But the more frequently we can make connection our first response, regardless of what our children have done, or whether or not we ourselves are in the river of well-being, the more we'll show our kids that they can count on us to offer solace, unconditional love, and support, even when they've acted in ways we don't like. Talk about fortifying and deepening a relationship! What's more, in strengthening your own relationship with your children, you'll be better equipping them to be good siblings, friends, and partners as they move toward adulthood. You'll be teaching by role modeling, guiding by what you do and not only by what you say. That's the relational benefit of connection: it teaches kids what it means to be in a relationship and to love, even when we're not happy with the choices made by the person we love.

What About Tantrums? Aren't We Supposed to Ignore Them?

When we teach parents about connecting and redirecting, one of the most common questions we hear is about tantrums. Usually someone in the audience will ask something like, "I thought we were supposed to ignore tantrums. Doesn't connecting with a kid when he's freaking out just give him attention? So doesn't that just reinforce the negative behavior?"

Our response to this question reveals another place where the No-Drama, Whole-Brain philosophy deviates from conventional approaches. Yes, there may be times when a child throws what we might call a strategic tantrum, when he's in control of himself and is willfully acting distressed to achieve a desired end: to get a toy he wants, to stay at the park longer, and so on. But with most children, and almost always with young children, strategic tantrums are much, much more the exception than the rule.

The vast majority of the time, a tantrum is evidence that a child's downstairs brain has hijacked his upstairs brain and left him legitimately and honestly out of control. Or, even if the child isn't fully dysregulated, he's enough out of sorts in his nervous system that he whines or doesn't have the capacity to be flexible and manage his feelings in that moment. And if a child is unable to regulate his emotions and actions, our response should be to offer help and emphasize comfort. We should be nurturing and empathic, and focus on connection. Whether he's out of sorts and just beginning to move down the road to distress or so upset that he's actually out of control, he *needs* us in this moment. We still need to set limits—we can't let a child, in his distress, yank down the curtains at the restaurant—but our objective in that moment is to comfort him and help him calm down so he can regain control of himself. Recall that chaos and losing control are signs of blocked integration, where the different parts of the brain are not working as a coordinated whole. And since connection creates integrative opportunities, connection becomes the way we comfort. Integration creates the ability to regulate emotions—and that's how we soothe our kids, helping them move from the chaos or rigidity of non-integrated states to the calmer and clearer harmony of integration and well-being.

So when parents ask for our opinion on tantrums, our response is that we need to completely reframe the way we think about the times our kids are the most upset and out of control. We suggest that parents view a tantrum not merely as an unpleasant experience they have to learn to get through, manage for their own benefit, or stop as soon as possible at all costs, but instead as a plea for help—as another opportunity to make a child feel safe and loved. It's a chance to soothe distress, to be a haven when an internal storm is raging, to practice moving from a state of dis-integration into a state of integration, through connection. That's why we call these moments of connection "integrative opportunities." Remember that a child's repeated experience of having her caregiver be emotionally responsive and attuned to her—connect with her—builds her brain's ability to self-regulate and self-soothe over time, leading to more independence and resilience.

So a No-Drama response to a tantrum begins with parental empathy. When we understand *why* children have tantrums—that their young, developing brains are subject to becoming dis-integrated as their big emotions take over—then we're going to offer a much more compassionate response when the screaming, yelling, and kicking begin. It doesn't mean we'll ever enjoy a child's tantrum—if you do, you might consider seeking professional help—but viewing it with empathy and compassion will lead to much greater calm and connection than seeing it as evidence of the child simply being difficult or manipulative or naughty.

That's why we're not at all fans of the conventional approach that calls for parents to completely ignore a tantrum. We agree with the notion that a tantrum is *not* the time to explain to a child that she's acting inappropriately. A child in the midst of a tantrum is not experiencing what is traditionally called a "teachable moment." But the moment can be transformed through connection into an integrative opportunity. Parents tend to overtalk in general when their kids are upset, and asking questions and trying to teach a lesson mid-tantrum can further escalate their emotions. Their nervous systems are already overloaded, and the more we talk, the more we flood their systems with additional sensory input.

But that fact doesn't at all logically lead to the conclusion that we should ignore our children when they're distraught. In fact, we're encouraging pretty much the opposite response. Ignoring a child in the midst of a tantrum is one of the worst things we can do, because *when a child is that upset, he's actually suffering.* He is miserable. The stress hormone cortisol is pumping through his body and washing over his brain, and he feels completely out of control of his emotions and impulses, unable to calm himself or express what he needs. That's suffering. *And just like our kids need us to be with them and provide reassurance and comfort when they're physically hurting, they need the same thing when they're suffering emotionally. They need us to be calm and loving and nurturing. They need us to connect.*

We know how unpleasant a tantrum can be. Believe us, we know. But here's what it really comes down to. What message do you want to send your children?

MESSAGE 1:

MESSAGE 2:

When you deliver this second message, you're not giving in. You're not being permissive. It doesn't mean you have to let a child harm himself, destroy things, or put others at risk. You can, and should, still set boundaries. You may even have to help him control his body or stop an impulse during a tantrum. (We'll offer specific suggestions for doing so in the coming chapters.) *But you set these limits while communicating your love and walking through the difficult moment with your child, always communicating, "I'm here."*

Of course we want the tantrum to resolve as quickly as possible, just like we want to get out of the dentist chair as soon as we can. It's simply not pleasant. But if you're working from a Whole-Brain perspective, the quickest ending to the tantrum is really not your primary goal. Rather, your first objective is to be emotionally responsive and present for your child. Your primary goal is to connect—which will offer all the short-term, long-term, and relational benefits we've been discussing. In other words, even though you want the tantrum to end as soon as possible, the larger goal of connecting actually gets you there a lot more efficiently in the short run, and achieves a whole lot more in the long run. You'll make things easier and less dramatic for both your child and yourself by providing empathy and your calm presence during a tantrum, and you'll build your child's capacity to handle himself better in the future, because emotional responsiveness strengthens the integrative connections in his brain that allow him to make better choices, control his body and emotions, and think about others.

How Do You Connect Without Spoiling a Child?

We've said that connection defuses conflict, builds a child's brain, and strengthens the parent-child relationship. One question parents often raise, though, has to do with a potential drawback of connecting before redirecting: "If I'm always connecting when my kids do something wrong, won't I spoil them? In other words, won't that reinforce the behavior that I'm trying to change?"

These reasonable questions are based on a misunderstanding, so let's take a few moments and discuss what spoiling is, and what it's not. Then we can be more clear on why connecting during discipline is quite different from spoiling a child.

Let's start with what spoiling is not. *Spoiling is not about how much love and time and attention you give your kids. You can't spoil your children by giving them too much of yourself.* In the same way, you can't spoil a baby by holding her too much or responding to her needs each time she expresses them. Parenting authorities at one time told parents not to pick up their babies too much for fear of spoiling them. We now know better. Responding to and soothing a child does not spoil her—but *not* responding to or soothing her creates a child who is insecurely attached and anxious. Nurturing your relationship with your child and giving her the consistent experiences that form the basis of her accurate belief that she's entitled to your love and affection is exactly what we *should* be doing. In other words, we want to let our kids know that they can count on getting their *needs* met.

Spoiling, on the other hand, occurs when parents (or other care-givers) create their child's world in such a way that the child feels a sense of entitlement about getting her way, about getting what she *wants* exactly when she wants it, and that everything should come easily to her and be done for her. We want our kids to expect that their *needs* can be understood and consistently met. But we don't want our kids to expect that their *desires and whims* will always be met. (To paraphrase the Rolling Stones, we want our kids to know they'll get what they need, even if they can't always get what they want!) And connecting when a child is upset or out of control is about meeting that child's needs, not giving in to what she wants.

The dictionary definition of "spoil" is "to ruin or do harm to the character or attitude by overindulgence or excessive praise." Spoiling can of course occur when we give our kids too much stuff, spend too much money on them, or say yes all the time. But it also occurs when we give children the sense that the world and people around them will serve their whims.

Is the current generation of parents more likely to spoil their kids than previous generations? Quite possibly. We see this most commonly when parents shelter their children from having to struggle at all. They overprotect them from disappointments or difficulties. Parents often confuse indulgence, on one hand, with love and connection, on the other. If parents themselves were raised by parents who weren't emo-tionally responsive and affectionate, they often experience a well-meaning desire to do things differently with their own kids. *The problem appears when they indulge their children by giving them more and more stuff, and sheltering them from struggles and sadness, instead of lavishly offering what their kids really need, and what really matters— their love and connection and attention and time—as their children struggle and face the frustrations that life inevitably brings.*

There's a reason we worry about spoiling our kids by giving them too much stuff. When kids are given whatever they want all the time, they lose opportunities to build resilience and learn important life lessons: about delaying gratification, about having to work for some-thing, about dealing with disappointment. Having a sense of entitlement, as opposed to an attitude of gratitude, can affect relationships in the future, when the entitled mind-set comes across to others.

We also want to give our children the gift of learning to work through difficult experiences. We're doing our child no favor when we find his unfinished homework on the kitchen table and complete it ourselves before running it up to school to protect him from facing the natural consequences of a late assignment. Or when we call another parent to ask for an invitation to a birthday party that our child caught wind of but was not invited to. These responses create an expectation in children that they'll experience a pain-free existence, and as a result, they may be unable to handle themselves when life doesn't turn out as they anticipated.

Another problematic result of spoiling is that it chooses immediate gratification—for both child and parent—over what's best for the child. Sometimes we overindulge or decide not to set a limit because it's easier in the moment. Saying yes to that second or third treat of the day might be easier in the short term because it avoids a meltdown. But what about tomorrow? Will treats be expected then as well? Remember, the brain makes associations from all of our experiences. Spoiling ultimately makes life harder on us as parents because we're constantly having to deal with the demands or the meltdowns that result when our kids don't get what they've come to expect: that they'll get their way all the time.

Spoiled children often grow up to be unhappy because people in the real world don't respond to their every whim. They have a harder time appreciating the smaller joys and the triumph of creating their own world if others have always done it for them. True confidence and competence come not from succeeding at getting what we want, but from our actual accomplishments and achieving mastery of something on our own. Further, if a child hasn't had practice dealing with the emotions that come with not getting what she wants and then adapting her attitude and comforting herself, it's going to be quite difficult to do so later when disappointments get bigger. (In Chapter 6, by the way, we'll discuss some strategies for dialing back the effects of spoiling if we've gotten into that unhelpful habit.)

What we're saying is that parents are right to worry about spoiling their kids. Overindulgence is unhelpful for children, unhelpful for parents, and unhelpful for the relationship. But spoiling has nothing to do with connecting with your child when he's upset or making bad choices. Remember, you can't spoil a child by giving him too much emotional connection, attention, physical affection, or love. When our children need us, we need to be there for them.

Connection, in other words, isn't about spoiling children, coddling them, or inhibiting their independence. When we call for connection, we're not endorsing what's become known as helicopter parenting, where parents hover over their children's lives, shielding them from all struggle and sadness. Connection isn't about rescuing kids from adversity. *Connection is about walking through the hard times with our children and being there for them when they're emotionally suffering, just like we would if they scraped their knee and were physically suffering.* In doing so, we're actually building independence, because when our children feel safe and connected, and when we've helped them build relational and emotional skills by disciplining from a Whole-Brain perspective, they'll feel more and more ready to take on whatever life throws their way.

You Can Connect While Also Setting Limits

So yes, as we discipline our children, we want to connect with them emotionally and make sure they know we're there for them when they're having a hard time. But no, this doesn't at all mean we should indulge their every whim. In fact, it would be not only indulgent but irresponsible if your child were crying and tantruming at the toy store because she didn't want to leave, and you allowed her to keep screaming and throwing anything she could get her hands on.

You're not doing a child any favors when you remove boundaries from her life. It doesn't feel good to her (or to you or the other people in the toy store) to allow her emotional explosion to go unfettered. When we talk about connecting with a child who's struggling to control herself, we don't mean you allow her to behave however she chooses. You wouldn't simply say "You seem upset" to a child as he hurls a Bart Simpson action figure toward a breakable Hello Kitty alarm clock. A more appropriate response would be to say something like, "I can see that you're upset and you're having a hard time stopping your body. I will help you." You might need to gently pick him up or guide him outside as you continue to connect—using empathy and physical touch, remembering that he's needing you—until he's calm. Once he's more in control of himself and in a state of mind that's receptive to learning, then you can discuss what happened with him.

Notice the difference in the two responses. One ("You seem upset") allows the child's impulses to hold everyone captive, leaving him unaware of what the limits are, and doesn't give him the experience of putting on the brakes when his desires are pressing the gas pedal. The other gives him practice at learning that there are limits on what he can and can't do. Kids need to feel that we care about what they're going through, but they also need us to provide rules and boundaries that allow them to know what's expected in a given environment.

INSTEAD OF LIMITLESS INDULGENCE...

SET LOVING LIMITS

When Dan's children were small he took them to a neighborhood park where he witnessed a four- or five-year-old boy being bossy and too rough with the children around him, some of them quite little. The boy's mother chose not to intervene, ostensibly because she'd "rather not solve his problems for him." Eventually another mom let her know that the boy was being rough and preventing children from using the slide, at which time his mother harshly reprimanded him from across the way: "Brian! Let those kids slide or we're going home!" In response, he told her that she was stupid and began throwing sand. She said, "OK, we're going," and began gathering up their things, but he refused to leave. The mom kept making threats but took no action. When Dan left with his kids ten minutes later, the mom and her son were still there.

This situation raises a question about what we mean when we talk about connecting. In this case, the issue at hand wasn't that the boy was upset and crying. He was still having a hard time regulating his impulses and handling the situation, but it was expressed more in stubborn and oppositional behavior. Still, connection was in order before his mother attempted to redirect him. When a child isn't overwhelmed by emotions but is simply making less-than-optimal decisions, connection might mean acknowledging how he's feeling in that moment. She could walk over and say, "It looks like you're having fun deciding who gets to use the slide. Tell me more about what you and your friends are doing here."

A simple statement like this, said in a tone that communicates interest and curiosity instead of judgment and anger, establishes an emotional connection between the two of them. The boy's mother can then more credibly follow up with her redirection, which might express the same sentiment she used earlier, but do so in a very different tone. Depending on her own personality and her son's temperament, she could say something like, "Hmmm. I just heard from another mom that some of the kids are wanting to use the slide, and that they're not liking how you're blocking it. The slide is for all the kids at the park. Do you have any ideas for how we can all share it?"

In a good moment, he might say something like, "I know! I'll go down and then run around and they can go down while I'm climbing back up." In a not-so-generous moment, he might refuse, at which time the mother might need to say, "If it's too hard to use the slide in a way that works for you *and* your friends, then we'll need to do something different, like throwing the Frisbee."

With these types of statements, the mother would be attuning to his emotional state, while still enforcing boundaries that teach that we need to be considerate of others. She could even give him a second chance if need be. But if he then refused to comply and began hurling more insults and more sand, she would have to follow through on the redirection she promised: "I can see you're really angry and disappointed about leaving the park. But we can't stay because you're having a hard time making good choices right now.

INSTEAD OF COMMANDING AND DEMANDING...

CONNECT WHILE SETTING LIMITS

Would you like to walk to the car? Or I can carry you there. It's your choice." Then she'd need to make it happen.

So yes, we want to always connect with our children emotionally.

But along with connecting, we must help kids make good choices and respect boundaries, as we clearly communicate and hold the limits. It's what children need, and even what they want, ultimately. Again, they don't feel good when their emotional states hold them and everyone else hostage. It leaves them on the chaos bank of the river, feeling out of control. We can help move their brains back toward a state of integration and move them back into the flow of the river by teaching them the rules that help them understand how the world and relationships work. Giving parental structure to our children's emotional lives actually gives them a sense of security and the freedom to feel.

We want our kids to learn that relationships flourish with respect, nurturing, warmth, consideration, cooperation, and compromise. So we want to interact with them from a perspective that emphasizes both connections and boundary setting. In other words, when we consistently pay attention to their internal world while also holding to standards about their behavior, these are the lessons they'll learn. From parental sensitivity and structure emerge a child's resourcefulness, resilience, and relational ability.

Ultimately, then, kids need us to set boundaries and communicate our expectations. But the key here is that all discipline should begin by nurturing our children and attuning to their internal world, allowing them to know that they are seen, heard, and loved by their parents—even when they've done something wrong. When children feel seen, safe, and soothed, they feel secure and they thrive. This is how we can value our children's minds while helping to shape and structure their behavior. We can help guide a behavioral change, teach a new skill, and impart an important way of approaching a problem, all while valuing a child's mind beneath the behavior. This is how we discipline, how we teach, while nurturing a child's sense of self and sense of connection to us. Then they'll interact with the world around them based on these beliefs and with these social and

WHAT CONNECTION LOOKS LIKE:

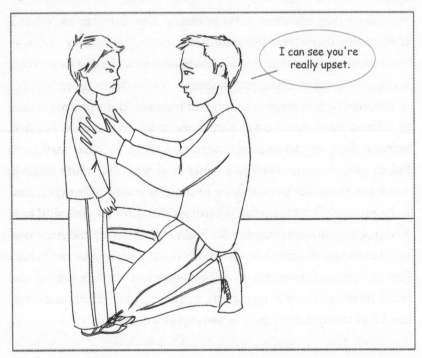

emotional skills, because their brains will be wired to expect that their needs will be met and that they are unconditionally loved.

So the next time one of your children loses control or does something that drives you completely crazy, remind yourself that a child's need to connect is greatest in times of high emotion. Yes, you'll need to address the behavior, to redirect and teach the lessons. But first, reframe those big feelings and recognize them for what they are: a bid for connection. *When your child is at his worst, that's when he needs you the most.* To connect is to share in your child's experience, to be present with him, to walk through this difficult time with him. In doing so you help integrate his brain and offer him the emotional regulation he's unable to access on his own. Then he can move back into the flow of the river of well-being. You will have helped him move from reactivity to receptivity, helped build his brain, and deepened and strengthened the relationship you two share.

CHAPTER 4

No-Drama Connection in Action

Tina and her family were eating dinner at home one night when she and her husband noticed that their six-year-old hadn't returned from the bathroom for several minutes. They found him playing on Tina's iPad in the living room. Here's how Tina tells the story:

> At first I was frustrated because my six-year-old had broken several of our rules. He had snuck away from the table, and he had played on the iPad without asking. He had also taken the iPad out of its protective case, which he knew he wasn't supposed to do. None of the infractions was significant. The problem was that he was disregarding the rules we had all agreed to.
>
> First, I thought about my son, and his temperament and developmental stage. As Dan and I have said several times now, context always has to be taken into consideration when deciding how to discipline. I knew that because my son is a sensitive and conscientious little guy, I probably wouldn't need to say much to discipline him.
>
> Scott and I sat on the couch next to him, and I simply said, in a curious tone, "What happened here?"

Immediately, my son's lower lip began to quiver, and tears pooled in his eyes. "I just wanted to try Minecraft!"

The nonverbal communication was a reflection of his inner conscience and his own discomfort, and the words were an admission of guilt. Implicit in his statement was the message, "I know I wasn't supposed to leave the table and get the iPad, but I just wanted to play so bad! My impulse was too strong." By this moment, in other words, I already knew that the redirection part of our conversation wasn't going to be too challenging. At other times it is, but not now, when there was already an awareness on his part.

Before redirecting, though, I wanted to meet him where he was, and to connect with him emotionally. I said, "You really are interested in that game, aren't you? You're curious about what the bigger boys are playing?"

Scott followed my lead and said something about how cool it is that the game allows you to create a whole world full of buildings and tunnels and animals.

Our son sheepishly looked up at us, moving his eyes from me to Scott, questioning whether things were really OK among us all. Then he nodded and gave us a soft smile.

With these few sentences and glances, connection had been established. Scott and I could then redirect. And again, knowing our son and recognizing where he was at this moment, the situation didn't require much from us. Scott simply asked, "But what about our rules?"

Here our son began to cry in earnest. Not much more needed to be said because the lesson had already been internalized.

I put my arm around him to comfort him. I said, "I know your choices tonight didn't follow our rules. Is there anything you'd do differently next time?"

He nodded as he cried, then promised to ask to be excused before leaving the table next time. We hugged, and then Scott asked him a Minecraft question, which led him to explain to his dad something about a trapdoor and a dungeon. As he became more animated, he moved past his guilt and his tears, and we all rejoined the rest of the family at the table. Connection had led to redirection, meaning not only that teaching could occur, but also that our son felt understood and loved.

Setting the Stage for Connection: Response Flexibility

In the previous chapter we discussed connection as the first step of the discipline process. Now we'll focus on what that actually looks like in action, recommending principles and strategies you can rely on when your child is upset or misbehaving. Sometimes connection is fairly simple, as it was here for Tina. Often, it's much more challenging.

As we discuss recommendations for connection, avoid the temptation to look for the formulaic one-size-fits-all technique that is supposed to apply in every situation. The following principles and strategies are extremely effective most of the time. But you should apply these approaches based on your own parenting style, the situation at hand, and your individual child's temperament. In other words, maintain response flexibility.

Response flexibility means just what it sounds like—being flexible about our response to a situation. It means pausing to think and to choose the best course of action. It lets us separate stimulus from response, so that our reaction doesn't immediately (and unintentionally) follow from a child's behavior or our own internal chaos. So when A happens, we don't just automatically do B; instead we consider B, C, or even a combo of D and E. Response flexibility creates a space in time and in our minds that enables a wide range of possibilities to be considered. As a result, we can just "be" with an experience, if only for a few seconds, and reflect before engaging the "do" circuitry of action.

Response flexibility helps you choose to be your wisest self possible in a difficult moment with your child, so that connection can occur. It's pretty much the opposite of autopilot discipline, where you apply a robotic one-size-fits-all approach to every scenario that arises. When we're flexible in our responses to our children's state of mind and their misbehavior, we allow ourselves to intentionally respond to a situation in the best way possible and provide our kids with what they need in the moment.

Depending on your child's infraction, this might require taking a moment to calm down. It's a good rule of thumb not to respond the nanosecond after you witness a misbehavior. We know you may *feel*, in the heat of the moment, like laying down the law, yelling that since your daughter pushed her brother into the pool, she's done swimming for the rest of the summer. (Aren't we ridiculous sometimes?) But if you can take a few seconds and allow yourself to calm down, rather than making a scene at the public pool and overshooting the discipline mark, you'll have a better chance at intentionally responding out of a calmer and more thoughtful part of yourself to what your child actually needs right then. (As a bonus, you can avoid being the subject of dinner conversations all around town that begin, "You should've seen this crazy lady at the pool today.")

At other times, response flexibility may lead you to decide to take a *firmer* stand on an issue than you normally might. If you notice signs that your eleven-year-old is taking less initiative with his responsibilities and his schoolwork, you might decide *not* to drive him back to school so he can retrieve the book he (again!) "has no idea how" he left in his locker. You would sincerely empathize with him and make sure to connect—"It's such a bummer that you forgot your book and won't have your assignment ready tomorrow"—but you'd allow him to experience the natural and logical fallout of his forgetfulness. Or maybe you *would* take him to get his book, because his personality or the context of the situation leads you to believe that approach would be best. That's the whole point. Response flexibility means you're making a point to *decide* how you want to respond to each situation that arises, rather than simply reacting without thinking about it.

Like so many aspects of parenting, response flexibility is funda-mentally about parenting intentionally. We're talking about remain-ing mindful of meeting the needs of your child—*this particular child*—in this particular moment. When that goal is central in your mind, connection will necessarily follow.

Now let's look at some specific ways you can use response flexibil-ity to connect with your kids when they're having a hard time han-dling themselves well or when they're making unwise decisions. We'll start by focusing on three No-Drama connection principles that set the stage and allow for connection between parent and child. Then we'll move to some more immediate in-the-moment connection strategies.

Connection Principle #1: Turn Down the Shark Music

If you've heard Dan speak, you may have seen him introduce the concept of shark music. Here's how he explains the idea:

> First, I ask the audience to monitor the response of their bod-ies and minds as I show them a thirty-second video.* On the screen, the audience sees what appears to be a beautiful forest. From the point of view of the person holding the camera, the audience sees a rustic trail and moves down that path toward a beautiful ocean. All the while, calm, classical-sounding piano music plays, communicating a sense of peace and seren-ity in an idyllic environment.
>
> I then stop the video and ask the audience to watch it again, explaining that I'm going to show them the exact same video, but this time different music will play in the background. The audience then sees the same images—the forest, the rustic trail, the ocean. But the soundtrack this time is dark and men-

---

* This video was originally produced by the Circle of Security Intervention Program. See their great work in the book *The Circle of Security Intervention* by Bert Powell et al. (New York: Guilford Press, 2013).

acing. It's like the famous theme music from the movie *Jaws*, and it completely colors the way the scene is perceived. The peaceful scene now looks threatening—who knows what might jump out?—and the path leads somewhere we're pretty sure we don't want to go. There's no telling what we'll find in the water at the end of the trail; based on the music, it's likely a shark. But despite our fear, the camera continues to approach the water.

The exact same images, but as the audience discovers, the experience drastically changes with different background music. One soundtrack leads to peace and serenity, the other to fear and dread.

It's the same when we interact with our children. We have to pay attention to our background music. "Shark music" takes us out of the present moment, causing us to practice fear-based parenting. Our attention is on whatever we are feeling reactive about. We worry about what's coming in the future, or we respond to something from the past. When we do so, we miss what's actually happening in the moment—what our children really need, and what they're actually communicating. As a result, we don't give them our best. Shark music, in other words, keeps us from parenting this individual child in this individual moment.

For instance, imagine that your fifth grader comes home with her first progress report, which shows that, since she was sick and missed a couple of days of class, her math average is lower than you'd expect. Without shark music playing in the background, you might just chalk this up to the absences, or to the more difficult subject matter in fifth grade. You'd take steps to make sure she understands the material now, and you might or might not decide to visit with her teacher. In other words, you'd approach the situation from a calm and rational perspective.

If, however, your daughter's older brother is a ninth grader who has shown himself to be less than responsible with his homework, and who is struggling with the basics of algebra, this prior experience might become shark music that plays in your mind as your daughter shows you her progress report. "Here we go again" might be the refrain that takes over your thoughts. So instead of responding to the situation as you normally would, asking your daughter how she feels about it and trying to figure out what's best for her, you think about your son's problems with algebra, and you overreact to your daughter's situation. You begin talking to her about consequences, and cutting back on after-school activities. If the shark music *really* gets to you, maybe you start lecturing about getting into good colleges, and the chain of events that leads from a couple of bad grades in fifth-grade math to problems in middle school and high school and finally to a slew of rejection letters from universities all across the country. Before you know it, your adorable ten-year-old has become a homeless woman pushing a shopping cart toward the cardboard box she lives in under the bridge down by the river—all because she got mixed up about which way the "greater than" symbol points!

The key to a No-Drama response, as is so often the case, is awareness. Once you *recognize* that shark music is blaring in your mind, you can shift your state of mind and stop parenting based on fear and on past experiences that don't apply to the current scenario you face. Instead, you can connect with your child, who might be feeling discouraged. You can give her what she needs in this moment: a parent who is fully present, parenting *only her* based *only on the actual facts of this particular situation*—not on past expectations or future fears. See illustrations on next page.

This isn't to say that we don't pay attention to patterns of behavior over time. We can also get trapped in states of denial where we over-contextualize behavior or explain away our kids' repeated struggles with all kinds of excuses that keep us from seeking intervention or from helping our children build the skills they need. You've met the parent who has a child who is never at fault and whom the parents never hold accountable. See illustrations on next page.

When the "excuse flavor of the week" becomes a pattern of parental response, then the parents are probably working from a different

INSTEAD OF HEARING SHARK MUSIC...

CONNECT WITH YOUR CHILD WHO NEEDS YOU

DENIAL CAN BE UGLY...

kind of shark music. It's similar to the parents whose children were medically vulnerable as babies, whose shark music now leads them to overdo for their kids, treating them as if they are still more fragile than they actually are.

The point is that shark music can prevent us from parenting intentionally and from being who our children need us to be at any given moment. It makes us reactive instead of receptive. Sometimes we're called to adjust our expectations and realize that our children just need more time for development to unfold; at other times we need to adjust our expectations and realize that our children are capable of *more* than we're asking of them, so we can challenge them to take more responsibility for their choices. At still other times we need to pay attention to our own needs, desires, and past experiences, any of which can override our ability to make good moment-by-moment decisions. The problem is that when we are reactive, we can't receive input from others, or demonstrate any response flexibility to consider the various options in our own mind. (If you'd like to go deeper with this concept, Dan covers it extensively in *Parenting from the Inside Out*, co-authored with Mary Hartzell.)

Ultimately, our job is to give unconditional love and calm presence to our kids even when they're at their worst. *Especially* when they're at their worst. That's how we stay receptive instead of going reactive. And the perspective we take on their behavior will necessarily affect how we respond to them. If we recognize them for the still-developing young people they are, with changing, changeable, complex young brains, then when they struggle or do something we don't like, we'll be better able to be receptive and hear the calming piano music. We'll therefore interact with them in a way that's more likely to lead to peace and serenity.

Shark music, on the other hand, will take us out of the present moment, and out of our right minds, as we become reactive. It will fuel our internal chaos and lead us to make all kinds of assumptions, to worry about all kinds of possibilities that simply shouldn't be considered in this particular situation. It might even lead us to automatically assume that our kids are "acting out" because they are selfish, lazy, spoiled, or whatever other label we choose. Then we'll respond not out of love and intentionality, but out of reactivity, anger, anxiety, drama, and fear.

So the next time you need to discipline, pause for just a second and listen for the soundtrack in your head. If you hear calm piano music and feel capable of offering a loving, objective, clearheaded response to the situation, then go ahead and offer just that kind of response. But if you notice the shark music, be very careful about what you do and say. Give yourself a minute—longer, if necessary—before responding. Then, when you feel yourself letting go of the fears, expectations, and bigger-than-necessary reactivity that keep you from looking at the situation for what it really is, you can respond. Simply by paying attention to whatever music is playing in the background of a disciplinary moment, you'll be much more capable of *responding flexibly* instead of *reacting rigidly or chaotically,* and offering your children what they need right then. Responding rather than reacting is the key.

Connection Principle #2: Chase the Why

One of the worst by-products of shark music is the parental tendency to make assumptions about what we perceive to be obvious. If a scary or emotionally charged soundtrack is clouding your mind when you interact with your children, you're not likely to be very objective about the reasons they're behaving the way they are. Instead, you're probably going to simply react based on information that might not be accurate at all. You'll assume there's a shark swimming in the water or a monster hiding behind the tree, even if there's not one.

When your kids are playing in the next room and you hear your younger child begin to cry, it may seem perfectly justifiable to march into the room, look at your older child, and demand, "What did you do this time?!" But when your younger child says, "No, Dad, I just fell and hurt my knee," you realize that what seemed obvious wasn't accurate at all, and that shark music has (once again) led you astray. Because your older child has played too rough in the past, you assumed that was the case this time.

Few parental actions will hinder connection faster than assuming the worst and reacting accordingly. So instead of making assumptions and operating on information that may be faulty, question what seems obvious. Become a detective. Put on your Sherlock Holmes hat. You know, Sherlock Holmes: the Arthur Conan Doyle character who declared, "It is a capital mistake to theorize before one has data. Insensibly one begins to twist facts to suit theories, instead of theories to suit facts."

When dealing with our children, it's dangerous to theorize before we have data. Instead, we need to be curious. We need to "chase the why."

Curiosity is the cornerstone of effective discipline. Before you ever respond to your child's behavior—especially when you don't like it—ask yourself a question: "I wonder why my child did that." Let this lead you to other questions: "What is she wanting here? Is she asking for something? Trying to discover something? What is she communicating?"

INSTEAD OF BLAMING AND CRITICIZING...

CHASE THE WHY

When a child acts in a way we don't like, the temptation will be to ask, "How could she *do* this?" Instead, chase the why. When you walk into the bathroom and see that your four-year-old has "decorated" the sink and mirror with wet toilet paper and a lipstick she found in a drawer, be curious. It's fine to be frustrated. But as quickly as possible, chase the why. Let your curiosity replace the frustration you feel. Talk to your daughter, and ask her what happened. Most likely you'll hear something that's totally plausible, at least from her perspective, and probably hilarious. The bad news is that you'll still have to clean up the mess (preferably with the help of your daughter). The good news is that you will have allowed your curiosity to lead you to a much more accurate—and fun, interesting, and honest—answer about your child's behavior.

The same would apply when your second grader's teacher calls to discuss certain "impulse control" problems your son is displaying. She tells you he's not respecting authority because he has begun making noises and inappropriate comments during class reading time. Your first reaction might be to initiate a "That's not the way we behave, mister" conversation with your son. But if you chase the why and ask him about his motivation, you might discover that "Truman thinks I'm funny when I do that, and now he lets me stand by him in the lunch line." You'll still need to do some redirecting, and work with your son on appropriate ways to navigate the difficult world of playground politics, but this way you'll be able to do so with much more accurate information about your son's emotional needs and what's actually driving his actions.

Chasing the why doesn't mean that we should necessarily ask our children "Why did you do that?" every time a disciplinary situation arises. In fact, that question may imply immediate judgment or disapproval, rather than curiosity. Further, sometimes children, especially young ones, may not know why they are upset or why they did what they did. Their personal insight and awareness of their own goals and motivations may not be very skilled yet. That's why we're not advising you to *ask* the why. We're recommending that you *chase*

INSTEAD OF BLAMING AND CRITICIZING...

CHASE THE WHY

the why. That's more about asking the why question in your own mind, allowing yourself to be curious, and wondering where your child is coming from in this moment.

Sometimes the behavior we want to address won't be as benign as lipstick decorations and potty humor. Sometimes our child will make decisions that lead to broken objects, bruised bodies, and damaged relationships. In these cases it's all the more important that we chase the why. We *need* to be curious about what drove our child to throw the screwdriver in anger, to strike another child, to spit out venomous words. It's not enough simply to address the behavior. Human behavior is purpose-driven most of the time. We need to know what's *behind* it, what's causing it. If we focus only on our child's behavior (her external world) and neglect the reasons behind that behavior (her internal world), then we'll concentrate only on the symptoms, not the cause that's producing them. And if we consider only the symptoms, we'll have to keep treating those symptoms over and over again.

But if we put on our Sherlock Holmes hat and chase the why, curiously looking for the root cause behind the behavior, we can more fully discover what's really going on with our child. We might find real reasons for concern that need to be addressed. Maybe we'll learn that our assumptions were false. Or maybe we'll discover that this "bad behavior" is an adaptive response to something that's too challenging for the child. Perhaps, for example, your child is faking an illness each day before PE class not because he's lazy or unmotivated or oppositional, but because that's his best strategy for dealing with the intense self-consciousness he feels when doing something athletic in front of his peers.

By wondering what our kids are trying to accomplish and by allowing them to explain a situation before we rush to judgment, we're able to gather actual data from their internal world, as opposed to simply reacting based on assumptions, faulty theories, or shark music. Plus, when we chase the why and connect first, we let our children know that we're on their side, that we're interested in their internal experience. We say to them, by the way we respond to each situation, that when we don't know what actually happened, we're going to give them the benefit of the doubt. Again, that doesn't mean turning a blind eye to misbehavior. It just means that we're looking to connect first, by asking questions and by being curious about what's behind the external behavior and what's happening inside our child.

Connection Principle #3: Think About the How

Listening for shark music and chasing the why are both principles that ask us to consider our own and our child's inner landscapes during a discipline moment. The third connection principle focuses on the way we actually interact with our kids. It challenges us to consider the way we talk to our children when they're having trouble managing themselves or making good decisions. *What* we say to our kids is of course important. But you know that just as important, if not more important, is *how* we say it.

Imagine that your three-year-old isn't getting into her car seat. Here are a few different *hows* for saying the exact same *what:*

- With eyes wide, big gestures, and a loud, angry tone of voice: *"Get in your car seat!"*
- With clenched teeth, squinted eyes, and a seething tone of voice: "Get in your car seat."
- With a relaxed face and a warm tone of voice: "Get in your car seat."
- With a wacky facial expression and a goofy voice: "Get in your car seat."

You get the idea. The *how* matters. At bedtime you might use a threat: "Get in bed now or you won't get any stories." Or you could say, "If you get in bed now, we'll have time to read. But if you don't get in bed right away, we'll run out of time and have to skip reading." The message is the same, but *how* it's communicated is very different. It has an entirely different feel. Both *how*s model ways of talking to others. Both set a boundary. Both deliver the same request. But they feel completely different.

It's the *how* that determines what our children feel about us and themselves, and what they learn about treating others. Plus, the *how* goes a long way toward determining their response in the moment, and how successful we'll be at helping produce an effective outcome that makes everyone happier. Children usually cooperate much more quickly if they feel connected to us, and when we engage them in a pleasant and playful exchange. It's the *how* that determines that. We can be much more effective disciplinarians if our *how* is respectful, playful, and calm.

So those are the three connection principles. By checking for shark music, chasing the why, and thinking about the how, we set the stage for connection. As a result, when our kids behave in ways we don't like, we have the opportunity to connect first, prioritizing the relationship and improving the odds of a successful disciplinary outcome. Now let's look at some specific connection strategies.

The No-Drama Connection Cycle

What does connection actually look like? What can we do to help our kids *feel felt* and know that we're with them, right in the middle of whatever they're going through, as we engage in the discipline process?

As always, the answer will change based on your individual child and your personal parenting style, but most often, connection comes down to a four-part, cyclical process. We call it the No-Drama connection cycle.

It won't always follow the exact same order, but for the most part, connecting with our children when they're upset or misbehaving involves these four strategies. The first: communicate comfort.

THE NO-DRAMA CONNECTION CYCLE

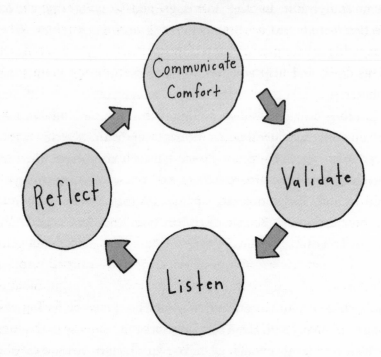

Connection Strategy #1: Communicate Comfort

Remember that sometimes your kids need your help calming down and making good choices. It's when their emotions get the best of them that we have the most discipline issues. And just as you'd hold and rock or pat a baby to calm her nervous system, you want to help your children calm down when they need it. Words are useful, especially when you're validating feelings. But most nurturing takes place nonverbally. We can communicate *so much* without ever talking.

The most powerful nonverbal response of all is one that you probably do automatically: you touch your child. You put your hand on her arm. You pull her close to you. You rub her back. You hold her hand. A loving touch—whether subtle, like the squeeze of a hand, or more demonstrative, like a full, warm embrace—has the power to quickly defuse a heated situation.

The reason is that when we feel someone touch us in a way that's nurturing and loving, feel-good hormones (like oxytocin) are released into our brain and body, and our levels of cortisol, a stress hormone, decrease. In other words, *giving your kids loving physical affection literally and beneficially alters their brain chemistry.* When your child (or your partner!) is feeling upset, a loving touch can calm things down and help you two connect, even during moments of high stress.

Touch is only one way we communicate with our children non-verbally. We're actually sending messages all the time, even when we never utter a word. Think about your typical body posture when you discipline your kids. Do you ever find yourself leaning over your child with an angry look on your face? Maybe you're saying, in a scary tone of voice, "Knock it off!" or "Stop that this instant!" This approach is essentially the opposite of connection, and it's not going to be very effective at calming your child. Your escalated response will intensify her emotions even further. Even if your intimidation results in your child *appearing* calm, she'll actually be feeling anything other than calm. Her heart will pound in response to the stress because she is afraid enough to shut down her emotions and hide her feelings in an attempt to keep you from becoming more angry.

Would you approach an upset animal in a similar fashion? If you had to interact with an angry-looking dog, would you approach it with an aggressive body posture and demand that the dog "knock it off and calm down"? That wouldn't be very smart, nor would it be effective. The reason is that it would communicate to the dog that you're a threat, and the dog would have no other option than to react, either by cowering or by fighting. So instead, we're taught to approach a dog by putting out the back of our hand, crouching down low, and speaking with a soft, reassuring voice. In doing so, our whole body communicates a message: "I'm not a threat." In response, the dog can relax, calm down, feel safe, and then approach and engage.

The same process occurs with people. When we feel threatened, our social engagement circuitry can't turn on. We have trouble engaging our upstairs brain, the part that is thoughtful, makes sound decisions, and has the capacity for empathy and regulating our emotions and body. Instead of calming down and making good decisions, we simply react. This reaction make sense, evolutionarily speaking. When the brain detects threat, the downstairs area immediately goes on alert and becomes highly activated. Functioning in this more primitive mode allows us to keep ourselves safe by being hypervigilant, by acting quickly without thinking, or by going into fight, flight, freeze, or faint mode.

It's the same with our kids. When emotions escalate and we respond by communicating threat—through the frustrated or angry look on our face, our mad tone of voice, our intimidating posture (hands on hips, finger wagging, leaning forward)—their innate biological response will be to activate their downstairs brain. However, when their caregivers communicate "I'm not a threat," then the reactive, fighting, act-before-thinking downstairs part of the brain quiets, and they can move into a mode of processing that allows them to handle themselves well.

So, how do we communicate "I'm not a threat" to our kids, even in the midst of escalating emotions? By connecting. One of the most effective and powerful ways to do this is to put your body in a posture that's the opposite of imposing and threatening. Lots of people talk about getting *at* a child's eye level, but one of the quickest ways to communicate safety and the absence of threat is to get *below* the child's eye level and put your body into a very relaxed position that communicates calm. You see other mammals doing this to send the message "I am not a threat to you. You don't need to fight me."

We recommend that you try this "below eye level" technique the next time your child is upset or out of control emotionally. Put your body in a chair, on a bed, or on the floor so that you are below your child's eye level. Whether you lean back or cross your legs or open your arms, just make sure that your body communicates comfort and safety. Your words and your body language combine to convey empathy and connection, telling your child, "I'm right here. I'll comfort you and help you." You'll comfort her nervous system and calm her down, just as you did by holding and rocking her when she was a baby and needed you.

COMMUNICATE COMFORT BY GETTING BELOW EYE LEVEL

We've been thrilled by how many of the parents to whom we've taught this technique report that this approach is "magic." They can't believe how quickly their children calm down. What amazes the parents just as much is that putting their bodies into this relaxed, non-threatening posture actually calms down *the parents themselves* as well. They report that this approach works better than anything else they've tried to keep themselves calm, and it leads to the best outcomes in how well *they* handle the high-stress situation. Obviously you can't get down on the ground if you're in the car or walking across the street, but you can use your tone of voice and posture, as well as your empathic words, to communicate the absence of threat, so you can connect with your child and produce a calm in both of you.

Nonverbal communication is so powerful. Your child's whole day can turn on something you're not even cognizant of, something that's not even said. Something as simple as your smile can soothe a disappointment and strengthen your bond. You know that moment: when your child does something she's excited about, like kicking a soccer goal or reciting a line in a play, and she looks for you in the crowd. Your eyes meet and you smile, and she knows that you're saying, "I saw that and I share your joy." That's what your nonverbal connection can do.

Or it can do just the opposite. Look at the pictures below, and notice what message these parents are sending. Without ever opening their mouths, each parent is saying plenty.

NONVERBALS ARE POWERFUL...

NONVERBAL MESSAGE: I'M EXASPERATED. YOU WEAR ME OUT. I CAN'T STAND YOU RIGHT NOW, AND I BLAME YOU FOR MAKING THINGS SO HARD ON ME.

NONVERBAL MESSAGE: I AM FURIOUS WITH YOU AND COULD EXPLODE AT ANY MOMENT. BE AFRAID, VERY AFRAID. THIS IS HOW PEOPLE ACT WHEN YOU DO SOMETHING WRONG.

NONVERBAL MESSAGE: YOU'D BETTER DO WHAT I SAY AND NOW! I DON'T CARE HOW YOU FEEL OR WHAT THE CIRCUMSTANCES ARE. POWER, CONTROL, AND AGGRESSION ARE HOW I GET WHAT I WANT.

The fact is that we send all kinds of messages, whether we intend to or not. And if we're not careful, our nonverbals can undermine the connection we're aiming for in a high-emotion disciplinary environment. Crossing the arms, shaking the head, rubbing the temples, rolling the eyes, a sarcastic wink at another adult in the room—even if our words are expressing interest in what our child is saying, there are plenty of ways our nonverbals betray us. And if our verbal and nonverbal messages contradict each other, our child will believe the nonverbal. That's why it's so important that we pay attention to what we're communicating without saying anything at all.

When we do, we'll be more likely to communicate the messages we *want* to communicate to our kids.

NONVERBALS ARE POWERFUL...

WHAT YOU'RE SHARING WITH ME RIGHT NOW IS CRUCIAL – MORE IMPORTANT THAN ANYTHING GOING ON AROUND US. EVEN MORE IMPORTANT THAN ANYTHING I WANT TO SAY.

I KNOW YOU HAD A HARD DAY AT SCHOOL, AND ALTHOUGH I DON'T HAVE JUST THE RIGHT WORDS TO SAY, I'LL ALWAYS BE HERE FOR YOU.

I THINK YOU'RE FANTASTIC, AND YOU FILL ME WITH JOY. I'M NOT EXACTLY HAPPY ABOUT THE DECISION YOU MADE, BUT I LOVE YOU EVEN WHEN YOU MESS UP.

We're not saying there won't be high-emotion disciplinary moments where you get completely exasperated with your kids. Or that they won't misread something you're communicating and get upset. Mistakes will be made on both sides of the relationship, of course. Likewise, sometimes you may decide it's appropriate to use nonverbal communication to help your kids monitor themselves and rein in their impulses when necessary. But the bottom line is that we can be intentional about the verbal *and* nonverbal messages we're sending, especially when we're trying to connect with our children in a difficult moment. Simply nodding, and being physically present, communicates care.

Connection Strategy #2: Validate, Validate, Validate

The key to connection when children are reactive or making bad choices is validation. In addition to communicating comfort, we need to let our kids know that we hear them. That we understand. That we get it. Whether or not we like the behavior that results from their feelings, we want them to feel acknowledged and sense that we're with them in the middle of all those big feelings.

Put differently, we want to *attune* to our children's inner subjective experiences, focusing our attention on how they are experiencing things from their point of view. Just as in a duet both instruments need to be tuned to each other to make good music, we need to tune our own emotional response to what's going on with our kids. We need to see their mind and recognize their internal state, then join with them in what we see and how we respond. In doing so, we join them in their emotional space. We deliver the message, "I get you. I see what you're feeling, and I acknowledge it. If I were in your shoes, and at your age, I might feel the same way." When kids receive this type of message from their parents, they "feel felt." They feel understood. They feel loved. And as a huge bonus, they can then begin to calm down and make better decisions, and hear the lessons you want to teach them.

Practically speaking, validation means resisting the temptation to deny or minimize what our kids are going through. When we validate their feelings we avoid saying things like, "Why are you throwing a fit about not having a playdate? You were at Carrie's all day yesterday!" We avoid pronouncing, "I know your brother tore your picture, but that's no reason to hit him! You can just make another one." We avoid declaring, "Stop worrying about it."

Think about it: how does it make *you* feel when you're upset, and maybe not handling yourself well, and someone tells you that you're "just tired," or that whatever's bothering you "isn't that big a deal" and you should "just calm down"? When we tell our kids how to feel—and how not to feel—we *in*validate their experiences.

Most of us know better than to directly tell our kids they shouldn't be upset. But when one of your children reacts intensely to something that doesn't go his way, do you ever immediately shut down that reaction? We don't mean to, but parents can often send the message that we think the way they feel about and experience a situation is ridiculous or not worthy of our acknowledgment. Or we inadvertently communicate that we don't want to interact with our kids or be with them when they have negative emotions. It's like saying, "I will not accept that you feel how you feel. I'm not interested in how you experience the world." It's a way of making a child feel invisible, unseen, and disconnected.

Instead, we want to communicate that we'll *always* be there for them, even at their absolute worst. We are willing to see them for whoever they are, whatever they may feel. We want to join with them where they are, and acknowledge what they're going through. To a young child we might say, "You really wanted to go to Mia's house today, didn't you? It's so disappointing that her mom had to cancel." Especially with older children, we might identify with what they're going through, letting them know that even though we're saying no to their behavior, we're saying yes to their feelings: "That made you so *mad* that Keith tore your picture, didn't it? I hate it when my stuff gets messed with, too. I don't blame you for being furious." Remember,

INSTEAD OF DISMISSING...

VALIDATE

the first response is to connect. Redirection will come, and you'll definitely want to address the behavioral response, but first we connect, which communicates comfort and almost always involves validation.

Usually validation is pretty simple. The main thing you need to do is simply identify the feeling at hand: "That really made you sad, didn't it?" or "I can see you feel left out," or even a more general "You're having a hard time." Identifying the emotion is an extremely powerful response when a child is upset because it offers two huge benefits. First, helping her feel understood calms her autonomic nervous system and helps soothe her big feelings, so she can begin to put the brakes on her desire to react and lash out. Second, it gives the child an emotional vocabulary and emotional intelligence, so she herself can recognize and name what she's feeling, which helps her understand her emotions and begin to regain control of herself so that redirection can occur. As we put it in the previous chapter, connection—in this case, through validation—helps move a child from reactivity to receptivity.

After acknowledging the feeling, the second part of validation is identifying with that emotion. For a child or an adult, it's extremely powerful to hear someone say, "I get you. I understand. I see why you feel this way." This kind of empathy disarms us. It relaxes our rigidity. It soothes our chaos. Even if an emotion seems ridiculous to you, don't forget that it's very real to your child, so you don't want to dismiss something that's important to her.

Tina recently received an email that reminded her that it's not only young children who need to be validated when they're upset. She heard from a mother in Australia who had listened to a radio program where Tina talked about the power of connection. Part of the mother's email went like this:

Right in the middle of listening I received a call from my nineteen-year-old daughter, who was having a meltdown. She was in pain from a physical therapy session, her bank account was in the negative, she didn't understand a lot of today's Business Law lecture, she was stressed about her exam tomorrow, and work wanted her to come in two hours early.

My first reaction was to say, "First-world problems. Suck it up, princess." But after hearing your interview, I realised that while indeed they were first-world problems, they were *her* first-world problems. So I said that I was sorry for her bad day, and did she need a mummy hug?

It made so much difference. I could hear her take a breath and relax. I told her I loved her, that her dad and I would fund her textbooks (which was why her bank account was in the negative), and that after her exam tomorrow I'd treat her to her favourite noodle soup at Bamboo Basket.

She was much more relaxed after the call, thanks to how I responded. So often we react harshly without realising the impact it may have. Even when our kids are mostly past the tantrum stage and we have a calm life with them, there's so many times throughout the day to put these ideas into practice.

Notice this mother's well-executed validation of her daughter's experience. She didn't invalidate her daughter's feelings by denying them, minimizing them, or blaming her. Instead, she acknowledged the bad day and asked whether she needed a hug. Her daughter's response was to take a deep breath and relax—not because her parents were going to help her financially, but because her feelings were acknowledged and identified. Because they were validated. Then the actual problems could be addressed.

So when your child is crying, raging, attacking a sibling, throwing a fit because his stuffed dog is too floppy and won't sit up properly, or demonstrating in any other way that he's incapable of making good decisions at that moment, validate the emotions behind the actions. Again, it might first be necessary to remove him from the situation. Validation doesn't mean allowing someone to get hurt or property to be destroyed. You're not endorsing bad behavior when you identify with your child's emotions. You're attuning to him. You're tuning your instrument to his, so that you two can create something beautiful together. You're meeting him where he is, looking for the meaning, the emotional undercurrent, behind his actions. You acknowledge and identify what he's feeling, and in doing so, you validate his experience.

Connection Strategy #3: Stop Talking and Listen

If you're like most of us, you talk way too much when you discipline. This response is actually funny if you think about it. Our child has gotten upset and made a bad decision, so we think, "I know. I'll lecture him. He'll calm down and make a better choice next time if I make him sit still and listen to me drone on and on about what he's done wrong." Want to turn your kids off, especially as they get older? Explain something, then keep making the same point over and over.

What's more, talking and talking to an emotionally activated child is not the least bit effective. When her emotions are exploding all over the place, one of the least effective things we can do is to talk at her, trying to get her to understand the logic of our position. It's just not helpful to say, "He didn't mean to hit you when he threw the ball; it was just an accident, so you don't have to get mad." It doesn't do any good to explain, "She can't invite everyone in your whole school to her party."

The problem with this logical appeal is that it assumes the child is capable of hearing and responding to reason at this moment. But remember, a child's brain is changing, developing. When she's hurt, angry, or disappointed, the logical part of her upstairs brain isn't fully functioning. That means a linguistic appeal to reason isn't usually going to be your best bet for helping her gain control over her emotions and calm herself.

In fact, talking often compounds the problem. We know, because we hear it from the kids we see in our offices. Sometimes they want to scream at their parents, "Please stop talking!" Especially when they're in trouble and already understand what they've done wrong. An upset child is already on sensory overload. And what does talking to him do? It further floods his senses, leaving him even more dysregulated, feeling even more overwhelmed, and much less able to learn or even hear you.

So we recommend that parents follow the kids' advice and stop talking so much. Communicate comfort and validate your child's feelings—"It really hurt that you didn't get invited, didn't it? I'd feel left out, too"—then close your mouth and listen. *Really* listen to what she's saying. Don't interpret what you hear too literally. If she says she's never going to get invited to another party, this isn't an invitation for you to disagree, or to challenge this absolute statement. Your job is to hear the feelings *within* the words. Recognize that she's saying, "I've really been thrown for a loop by this. I didn't get invited, and now I'm afraid about what this says about my social standing with all of my friends."

Clue in and chase the why as to what's really going on inside your child. Focus on her emotions, letting go of the shark music that prevents you from being fully present with her in the moment. No matter how strong your desire, avoid the temptation to argue with your child, lecture her, defend yourself, or tell her to stop feeling that way. Now's not the time to teach or explain. Now is the time to listen, just sitting with your child and giving her time to express herself.

Connection Strategy #4: Reflect What You Hear

With the first three strategies of the No-Drama connection cycle, we communicate comfort, we validate feelings, and we listen. The fourth step is to reflect back to our children what they've said, letting them know we've heard them. Reflecting their feelings returns us to the first strategy, since we're again communicating comfort, which can lead us through the cycle once more.

Reflecting what we hear is similar to the second step, but it differs from validation in that now we focus specifically on what our children have actually told us. The validation stage is all about recognizing emotions and empathizing with our kids. We say something like, "I can tell how mad you are." But when we reflect our children's feelings, we essentially communicate back to them *what they have told us.* Handled sensitively, this allows a child to feel heard and understood. As we said, it's extraordinarily calming, even healing, to feel understood. When you let your child know that you really grasp what he's telling you—by telling her, "I hear what you're saying; you really hated it when I told you we had to leave the party," or "No wonder that made you mad; I'd feel angry, too"—you take a huge step toward defusing the high emotions at play.

Be careful, though, with how you reflect feelings. You don't want to take one of your child's short-term, temporary emotions and turn it into something bigger and more permanent than it really is. Let's say, for example, your six-year-old becomes so upset about her big brother's constant teasing that she begins yelling, over and over again, "You're so stupid and I hate you!" Right there in your backyard, with the neighbors hearing it all (thank goodness Mr. Patel is mowing his lawn!), she repeats the refrain nonstop, seemingly dozens of times, until she finally falls into your arms, crying uncontrollably.

So you initiate the connection cycle. You communicate comfort, conveying your compassion by getting below her eye level, holding her, rubbing her back, and making empathic facial expressions. You validate her experience: "I know, honey, I know. You're really upset." You listen to her feelings, then you reflect back to her what you're hearing: "You're just so angry, aren't you?" Her response might be a return to yelling: "Yes, and I hate *Jimmy*!" (with her brother's name drawn out into another scream).

Now comes the tricky part. You want to reflect for her what she's feeling, but you don't want to reinforce this narrative in her mind that she actually hates her brother. A situation like this calls for some careful tiptoeing, so that you can be honest with your daughter and help her better understand her feelings but keep her from solidifying her momentary emotions into longer-lasting perceptions. So you might say something like, "I don't blame you for being so mad. I hate it when people tease me like that, too. I know you love Jimmy, and that you two were having so much fun together just a few minutes ago, when you were playing with the wagon. But you're pretty mad at him right now, aren't you?" The goal with this type of reflecting is to

THE NO-DRAMA CONNECTION CYCLE

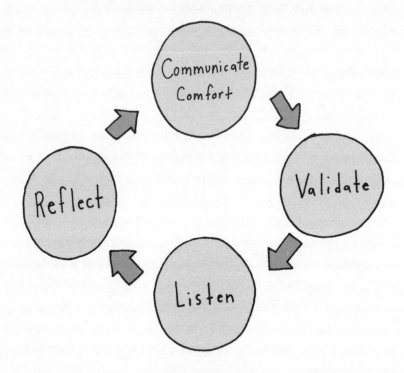

make sure your child comprehends that you understand her experience, and in doing so to soothe her big emotions and help calm her inner chaos, so that she can move back into the center of her river of well-being. But you don't want to allow a feeling that's simply a momentary *state*—her anger with her brother—to be perceived in her mind as a permanent *trait* that's an inherent part of their relationship. That's why you give her perspective and remind her of the fun she and her brother were having with the wagon.

One other advantage that comes with reflecting our children's feelings is that it communicates that they have not only our love, but our attention. Parents sometimes assume that it's bad when a child seeks our attention. They'll say, "He's just trying to get my attention." The problem with this perspective is that it presumes it's somehow not okay for a child to want his parents to notice him and pay attention to what he's doing. *In reality, though, attention-seeking behavior is not only completely developmentally appropriate, it's actually relational. Attention is a need of all children everywhere.* In fact, brain imaging studies show that the experience of physical pain and the experience of relational pain, like rejection, look very similar in terms of location of brain activity. So when we give our kids attention and focus on what they're doing and feeling, we meet an important relational and emotional need, and they deeply feel connected and comforted. Remember, there are plenty of ways to spoil children—by giving them too many *things*, by rescuing them from every challenge, by never allowing them to deal with defeat and disappointment—but we can never spoil them by giving them too much of our love and attention.

That's what the connection cycle does: it lets us communicate to our kids that we love them, that we see them, and that we are with them no matter how they behave. When we turn down the shark music, chase the why, and think about the how, we can communicate comfort, validate, listen to and reflect feelings, and support our kids in ways that create the kind of connection that clearly communicates our love and prepares them for redirection.

CHAPTER 5

1-2-3 Discipline:
Redirecting for Today, and for Tomorrow

Roger was working in his garage when his six-year-old, Katie, stormed outside, angrily calling out, "*Dad!* Can you do something about Allie?" Roger soon learned that Katie was upset because her friend Gina, who had come over for a playdate, had become completely enamored with Katie's nine-year-old sister, Allie. For her part, Allie was apparently happy to monopolize the playdate, leaving her younger sister feeling left out.

In addressing the situation with his older daughter, Roger saw various alternatives. One would be simply to tell Allie she needed to give Katie and Gina some time by themselves, since that was the plan for the playdate, after all. There would be nothing wrong with this approach, but by making the call and imposing his own agenda on the situation, Roger would bypass the important process that would allow Allie to use her upstairs brain.

So instead, he went into the house, called his older daughter aside, and simply initiated a brief conversation. They sat on the couch, and he put his arm around her. Considering Allie's personality and temperament, he decided to begin with a simple question:

ROGER: Gina's having fun playing with you, and you're really good with younger kids. But I'm wondering if you noticed that Katie's not too happy about Gina giving you all of her attention.

ALLIE: [Defensively, sitting up and turning toward her father] Dad, I'm not even doing anything mean. We're just listening to music.

ROGER: I didn't say you're doing anything wrong. I'm asking whether you've noticed how Katie is feeling right now.

ALLIE: Yeah, but that's not my fault!

ROGER: Sweetheart, I totally agree that it's not your fault. Listen to my question: do you see that Katie isn't happy? I'm asking whether you've noticed.

ALLIE: I guess.

In that one admission, we see evidence that Allie's upstairs brain had become engaged in the conversation, if only a little. She was actually beginning to listen and think about what her father was saying. At this point Roger could target which part of the upstairs brain he wanted to appeal to and exercise. Not by telling Allie what she should think or feel, but by asking her to consider the situation for herself, and to pay attention to what someone else was experiencing.

ROGER: Why do you think she might be upset?

ALLIE: I guess because she wants Gina all to herself. But that girl came into my room! I didn't even ask her to.

ROGER: I know. And you may be right that Katie wants Gina all to herself. But do you think that's it, exactly? If she were standing here and told us how she felt, what would she say?

ALLIE: That it's her playdate, not mine.

ROGER: That's probably pretty close. Would she have a point?

ALLIE: I just don't see why we can't all listen to music together. Seriously, Dad.

ROGER: I get it. I might even agree with you. But what would Katie say to that?

ALLIE: That when we're all together Gina just wants to play with me?

And with that question the empathy broke through. It was only an emerging awareness; we can't expect a huge Lifetime movie moment where a nine-year-old girl is moved to tears out of her compassion for her little sister's emotional pain. But it was a start. Allie was, at the very least, consciously beginning to consider the feelings of her sister (which, if you have young children, you know is no small parental victory). From there, Roger could direct the conversation so that Allie would think more explicitly about Katie's feelings. Then he could ask for Allie's help in coming up with a plan for handling the situation—"Maybe we listen to one more song, then I get ready for my slumber party?"—and in so doing he would further engage her upstairs brain by having her plan and problem-solve.

Initiating a redirection conversation like this won't always be suc-

INSTEAD OF COMMANDING AND DEMANDING...

ENGAGE THE UPSTAIRS BRAIN

cessful. There will be times when a child will be unwilling (or even unable) to see a different perspective, to listen and consider the feelings of others. Roger might end up simply telling Allie she needs to find something else to do, just as Liz had to make the call when her daughter wouldn't give in about who was going to drive her to school. Or maybe he could play a game with all three girls, making sure everyone feels included.

But notice that when he needed to redirect, Roger didn't immediately impose his own sense of justice on the situation. By facilitating empathy and problem solving, he gave his daughter a chance to exercise her upstairs brain. The more we give kids the opportunity to consider not only their own desires, but also the desires of others, and practice making good choices that positively impact the people around them, the better they'll be at doing so. Does a conversation like this one between Roger and Allie take longer than simply separating the girls? Of course. Is it harder to do? Yes, probably. But is collaborative and respectful redirection worth the effort and extra time? No question about it. And as it becomes your default, it actually makes things easier on you and your entire family, since there will be fewer battles, and you'll be building your child's brain in such a way that less and less often will you even have to address misbehavior.

1-2-3 Discipline

In this chapter we want to take a closer look at the concept of redirection, which is actually what most people mean when they think of discipline. Redirection is how we respond when our kids do something we don't like, such as throwing something in anger, or when they're not doing something we want them to do, like brush their teeth and get ready for bed. After we've connected, how do we address uncooperative or reactive kids, redirecting them toward using their upstairs brain so they can make more appropriate decisions that become second nature over time?

As we've said, No-Drama Discipline is about connecting and being emotionally responsive to our children, while aiming for the short-term goal of gaining cooperation now, as well as the long-term goal of building our child's brain. A simple way to think about redirection is to take a 1-2-3 approach that focuses on one definition, two principles, and three desired outcomes. You don't need to memorize every detail of the approach (especially since we've given you a handy Refrigerator Sheet at the back of the book). Just use it as an organizing framework to help you focus on what's important when it comes time to redirect your kids.

One Definition

The place to begin when thinking about redirecting our kids toward better behavior is with the definition of discipline. When our children make unwise decisions or can't manage their emotions, we need to remember that *discipline is about teaching.* If we forget this simple truth, we'll go off course. If discipline becomes about punishment, for example, we can miss the opportunity to teach. By focusing on the consequences of misbehavior, we limit the opportunity for children to experience the physiological and emotional workings of their inner compass.

One mom told us the story of finding a small box of crayons when she and her six-year-old were cleaning her daughter's room. They had been shopping for school supplies a few days before, and her daughter had fallen in love with these particular crayons. The mother had not bought the crayons, but her daughter had slipped them into her pocket anyway.

The mother said that when she found the crayons she decided to ask her daughter about them directly. When the little girl saw the crayons in her mom's cupped hand and the mother's look of confusion, her eyes got wide and full of fear and guilt. In a moment like this, the parental response is going to largely determine what a child takes away from the experience. As we explained in Chapter 1, if the parent's focus is on consequences or punishment, and she immediately yells, spanks, sends the child to her room, or takes away an upcoming opportunity she's excited about, then the child's focus will immediately shift. Instead of having her attention on that "uh-oh" feeling bubbling up inside of her, or instead of thinking about the decision she made when she took the crayons from the store, all of her attention will focus on how mean or scary her parent is for punishing her in this way. She may even feel like a victim, who is somehow retroactively justified in swiping the crayons.

Instead, this mom offered a disciplinary approach focused on teaching rather than immediate consequences. She gave her daughter time to sit with and be aware of that uncomfortable, valuable, natural guilt she was feeling as a result of having taken something that wasn't hers. Yes, guilt can even be healthy. It is evidence of a healthy conscience! And it can shape future behavior.

When the mother talked to her daughter, she knelt down (getting below eye level, as we discussed a few pages ago), and an endearing conversation ensued during which the six-year-old at first denied taking the crayons, then said she didn't remember, and then, with the mom patiently waiting, eventually explained that her mother had nothing to worry about, because "I waited till the saleslady with the big hair wasn't looking" to put the crayons in the pocket of her shorts. At this point the mother asked lots of questions that encouraged her daughter to think through concepts she hadn't yet considered: "Do you know what taking something that doesn't belong to you is called?" "Is stealing against the law?" "Did you know that the woman with the big hair in the store spent her money to buy those crayons so she could put them in her store?"

In response, the daughter dropped her head further; her bottom lip started to come forward, and big tears began to fall. She obviously felt bad about what she had done. As she quietly cried, the mom pulled her close, not distracting her or stopping the process of what was already happening naturally, but joining with her as she said, "You're feeling bad about it." The daughter nodded, and the tears continued. The mom could comfort and be with her daughter in this beautiful moment where the discipline process continued naturally without the mom even doing or saying anything. The mother held her and allowed her to cry and to feel, and after a couple of minutes she helped wipe away the tears and encouraged her daughter to take a deep breath. Then they continued their conversation briefly, talking about honesty, about respecting others' property, and about doing the right thing, even when it's hard.

By initiating this collaborative, reflective dialogue and allowing discipline to naturally arise simply by orienting her daughter's attention to the internal guilt she was already feeling, rather than just laying out instant consequences, the mother allowed her daughter to give her upstairs brain some exercise by considering her actions and how they affected others, and by learning some basic lessons about ethics and morality. Then they made plans for how best to return the crayons to "the saleslady with the big hair."

INSTEAD OF IMMEDIATELY GIVING CONSEQUENCES...

INITIATE A CONVERSATION

No-Drama Discipline is all about teaching, and that's what this mother focused on. She allowed her daughter to thoughtfully experience the feelings and thoughts associated with her decision to take the crayons. By allowing the child's own internal experience to remain at the forefront of her mind—rather than shifting the emotions into anger over a punishment handed out—she allowed her daughter's brain not only to become aware of that inner discomfort, but also to link it to the experience of making poor choices, in this case, stealing. Again, being punitive or doling out consequences, especially when we're angry and reactive, can be counterproductive because it distracts our children from the physiological and emotional messages of their own conscience, which is a powerful force in developing self-discipline.

Remember, neurons that wire together fire together. And we want our kids to experience the natural linkage between making a bad decision in one moment, then feeling guilty and ill at ease the next. Because the brain is driven to avoid experiences that produce negative sensations, the aversive feelings that naturally arise within a child when she does something that violates her inner conscience can be very fleeting in her conscious mind. But when we help her become aware of these sensations and emotions, they can become the important basis for ethics and self-control. *This self-regulation or executive function that develops can then engage even when her parent isn't there, or when no one is looking.* This is how she internalizes the lesson on a synaptic level. Our own nervous systems can become our very best guides!

Different disciplinary situations will obviously call for different parental responses. This mother responded based on what lesson her daughter needed in that particular moment. In other circumstances she might respond differently. The point is simply that once we've connected with our kids in a disciplinary moment and it's time to redirect, we've got to keep in mind the importance of awareness and helping the brain learn. Reflection with a child helps him become aware of what's happening internally, and that optimizes learning.

When we keep in mind the definition of discipline, we realize that sharing awareness helps learning occur. Discipline is all about teaching to optimize learning.

Two Principles

We also want to follow two main principles when redirecting our children, allowing those two principles to guide whatever we do. These principles, along with the specific strategies that follow from them, encourage cooperation from kids and make life easier for adults and kids alike.

Principle #1: Wait Until Your Child Is Ready

Remember what we said in Chapter 3: connection moves a child from reactivity to receptivity. So once you've connected and allowed your child to come to a place where he's ready to listen and use his upstairs brain, then it's time to redirect. Not before. One of the most self-defeating parenting recommendations we hear from time to time goes something like this: "When a child misbehaves, it's important that you address the behavior *right away*. Otherwise, they won't understand why they are being disciplined."

We actually don't think this is bad advice if you are running a behavioral conditioning lab with animals. For mice, or even dogs, it's good advice. For human beings, not so much. The fact is that there are times when it does make sense to address misbehavior right away. However, it's frequently the case that the absolute *worst* time to address a misbehavior is immediately after it's occurred.

The reason is simple. Misbehavior often happens because a child isn't able to regulate his big feelings. And when his emotions are dysregulated, his upstairs brain has gone off-line. It's temporarily out of order, meaning he's not able to accomplish the tasks his upstairs brain is responsible for: making good decisions, thinking about others, considering consequences, balancing his emotions and body, and being a receptive learner. So yes, we do recommend that you address a behavioral issue fairly soon when possible, but only when your child is in a calm and receptive state of mind—even if you need to wait. Even children as young as three can remember what happened in recent history, including the day before. You can begin that conversation by saying, "I'd like to talk about what happened yesterday at bedtime. That didn't go so well, did it?" Waiting for the right time is essential when it comes to teaching effectively.

So let's go back to the suggestion we made in Chapter 4. Once you've connected, and you're wondering whether it's time to move into the redirection phase, ask yourself one simple question: "Is my child ready? Ready to hear, ready to learn, ready to understand?" If the answer is no, then there's no reason to try to redirect in that moment. Most likely, more connection is called for. Or, especially for older kids, you may just need to give them some time and space before they'll be ready to hear you.

When we talk to educators, we often explain that there's an optimal window, or sweet spot, for teaching. If students' nervous systems are what we call *underaroused*—because they are sleepy, bored, or checked out for some other reason—then they are in an unreceptive state, meaning the students won't be able to learn effectively. And the opposite is just as bad. If students' nervous systems are *overaroused*—meaning they feel anxious or stressed-out, or their bodies are hyperactive with lots of motor activity and movement—that also produces an unreceptive state when it's difficult for them to learn. Instead, we need to create an environment that helps them move into a state of mind that's calm, alert, and receptive. That's the sweet spot where learning really takes place. That's the moment they're ready to learn.

It's the same with our kids. When their nervous systems are under- or overaroused, they won't be nearly as receptive to what we want to teach them. So when we discipline, we want to wait until they are calm, alert, and receptive. Ask yourself: "Is my child ready?" Even after you've connected and soothed your child's negative state, it still might be best to wait for a time later in the day or even the next day to find a better moment for the explicit teaching and redirection. You can even say, "I'd like to wait until we're really able to talk and listen to each other. We'll come back and talk about it in a while."

As a side note, just as it's important to ask, "Is my child ready?" it's also important to ask yourself, "Am I ready?" If you are in a reactive state of mind, it's best to wait to have the conversation. You can't be an effective teacher if you're not in a calm and collected state. If you're too upset to remain in control, you're likely to approach the whole interaction in a way that's counterproductive to your goals of teaching and building connection. In that case, it's often better to say something like, "I'm too angry to have a helpful conversation right now, so I'm going to take some time to calm down, and then we'll talk in a bit." Then, once you are *both* ready, discipline will be more effective and feel better to both of you.

Principle #2: Be Consistent, but Not Rigid

There's no question about it: consistency is crucial when it comes to raising and disciplining our children. Many parents we see in our offices realize that they need to work on being more consistent with their kids—whether it's with bedtimes, limiting junk food or media, or just in general. But there are other parents who place such a high priority on consistency that it becomes a rigidity that's not good for their kids, themselves, or the parent-child relationship.

Let's get clear on the difference between the two terms. *Consistency* means working from a reliable and coherent philosophy so that our kids know what we expect of them and what they should expect from us. *Rigidity,* on the other hand, means maintaining an unswerving devotion to rules we've set up, sometimes without having even thought them through, or without changing them as our kids develop. As parents, we want to be consistent, but not rigid.

Kids obviously need consistency. They need to know what our expectations are, and how we will respond if they break (or even bend) agreed-upon rules. Your reliability teaches them about what to expect in their world. More than that, it helps them feel safe; they know they can count on you to be constant and steady, even when their internal or external world is chaotic. This kind of predictable, sensitive, attuned care is actually what builds secure attachment. It lets us provide our kids with what's called "safe containment," since they have a secure base and clear boundaries to help guide them when their emotions are exploding. Limits you set are like the guardrails on the Golden Gate Bridge. For a child, living without clear boundaries is as anxiety-provoking as driving over that bridge without guardrails to stop you from plummeting into San Francisco Bay.

But rigidity is not about safety or reliability; it's about stubbornness. It keeps parents from compromising when necessary, or looking at context and the intention behind a behavior, or recognizing the moments when it's reasonable to make an exception.

One of the main reasons parents become rigid with their children is because they are practicing a form of *fear-based parenting*. They worry that if they ever give in and allow a soft drink at one meal, they'll create a slippery slope and their kids will be drinking Mountain Dew for breakfast, lunch, and dinner for the rest of their lives. So they stick to their guns and deny the soft drink.

Or their six-year-old has a nightmare and wants to climb into bed with them because he's scared, but they worry that they'll be setting a dangerous precedent. They say, "We don't want him to develop bad sleep habits. If we don't nip it in the bud right now, he'll be a bad sleeper his whole childhood." So they stick to their guns and dutifully send him back to his bed.

We understand the fear. We've felt it ourselves in regard to our own kids. And we agree that parents should definitely remain aware of whatever patterns they are setting up for their children. That's why consistency is so important.

But when fear-based parenting leads us to believe that we can *never* make an exception about a treat—or that we can't comfort or nurture our frightened child in the middle of the night without damning him to a life of sleeplessness—then we've moved into rigidity. That's parenting based on fear, not on what our child needs in that particular moment. That's parenting with a goal of reducing *our own anxiety and fears*, rather than what will best teach our child's emerging mind and mold the developing brain.

So how do we maintain consistency without crossing over to fear-based rigidity? Well, let's start by acknowledging that there are some non-negotiables. For instance, under no circumstances can you let your toddler run through a busy parking lot, or your school-age child swim without supervision, or your teenager get into a car with a driver who's been drinking. Physical safety is non-negotiable.

However, that doesn't mean you can't ever make exceptions, or even turn a blind eye from time to time when your child misbehaves. For instance, if you have a rule about no technology at the dinner table, but your four-year-old has just received a new electronic puzzle game that he'll play with quietly while you have dinner with another couple, that might be a good time to make an exception to your rule. Or if your daughter has promised that she'll finish her homework

RIGID

CONSISTENT BUT FLEXIBLE

before dinner but her grandparents show up to take her on an outing, you might negotiate a new deal with her.

The goal, in other words, is to maintain a *consistent but flexible* approach with your kids, so that they know what to expect from you, but they also know that at times you will thoughtfully consider all the factors involved. It goes back to what we talked about in the previous chapter: response flexibility. We want to intentionally respond to a situation in a way that considers what works best for our child and for our family, even if that means making an exception to our normal rules and expectations.

The question when it comes to consistent versus rigid discipline is what we're hoping to accomplish. Again, what do we want to teach? Under normal circumstances we want to consistently maintain our rules and expectations. But we want to avoid being rigid, ignoring context and thus missing out on the chance to teach the lessons we want to teach. Sometimes when we discipline, we need to look for other ways to accomplish our goals, so we can more effectively teach what we want our kids to learn.

At times, for example, you might try a "do-over." Instead of immediately offering a punishment for speaking disrespectfully, you can say something like, "I bet if you tried again, you could come up with a more respectful way to say that." Do-overs allow a child a second chance to handle a situation well. It gives them practice doing the right thing. You're still consistently maintaining your expectations, but you're doing so in a way that's often much more beneficial than a rigidly imposed, unrelated consequence.

After all, skill development is a huge part of what discipline is all about. And that requires repeated guidance and coaching. If you were coaching your child's soccer team and she was having trouble kicking the ball straight, you wouldn't give her consequences for every time she shanked it. Instead, you'd give her more practice, so that she gets better and better at kicking it where she wants it to go. You'd want her to have a clear, familiar feeling of what it's like to hit the ball square and watch it sail into the goal. In the same way, when our kids behave in ways that don't meet the expectations we've set up, sometimes the best thing we can do is to have them practice behaving in ways that do meet our expectations.

Another way to encourage skill building is to have your child come up with a creative response. As much as we wish it did, saying "I'm sorry" doesn't actually fix the broken fairy wand that was thrown in anger. An apology note and using allowance money to buy a new wand might teach more and help develop skills related to decision making and empathy.

The point is that in your efforts to build skills, you can still be consistent while remaining flexible and open to other alternatives. As kids learn about right and wrong, they also need to learn that life is not just about external reward and punishment. Flexibility, problem solving, considering context, and fixing our mistakes are also important. Most important is for children to understand the lesson at hand with as much personal insight as they are developmentally capable

INSTEAD OF RIGIDLY COMMANDING AND DEMANDING...

GIVE HIM PRACTICE DOING THE RIGHT THING

of, and to empathize with anybody they've hurt, then figure out how to respond to the situation and prevent it in the future.

In other words, there's a lot about morality that we want to teach our kids in addition to knowing right from wrong. We don't want to be their traffic cop, following them around telling them when to stop and when to go, and giving them tickets when they break the law. Wouldn't it be much better to teach them how to drive responsibly, and give them the skills, tools, and practice to make good decisions on their own? To do this successfully, sometimes we need to be open to seeing the gray areas, not just the black and white. We need to make decisions based not on an arbitrary rule we've previously set down, but on what's best for our kids and our family right now, in this particular situation. Consistent, yes, but not rigid.

Three Mindsight Outcomes

So 1-2-3 discipline focuses on *one* definition (teaching) and *two* principles (wait until your child is ready, and be consistent but not rigid). Now let's look at the *three* outcomes we're looking to achieve when we redirect.

If you've read *The Whole-Brain Child*, you're already familiar with the term "mindsight," which Dan coined and discusses at length in his books *Mindsight* and *Brainstorm*. Explained most simply, mindsight is the ability to see our own mind, as well as the mind of another. It allows us to develop meaningful relationships while also maintaining a healthy and independent sense of self. When we ask our children to consider their own feelings (using personal *insight*) while also imagining how someone else might experience a particular situation (using *empathy*), we are helping them develop mindsight.

Mindsight also involves the process of integration, which we discussed earlier. You'll remember that integration occurs when separate things become linked—like the right and left sides of the brain, or two people in a relationship. When integration does not occur,

INSIGHT +
EMPATHY =
MINDSIGHT

chaos or rigidity results. So when a relationship experiences an inevitable rupture in how we honor each other's differences, or when we don't link compassionately to each other, that's a break in integration. One example of creating integration is when we *repair* such a rupture. If you find that chaos or rigidity is popping up in your connection with your kids, repair is in order. We can take steps to repair the situation and make things right when we've made a bad decision or hurt someone with our words or actions. Let's discuss each of these outcomes (insight, empathy, and integration/repair) individually.

Outcome #1: Insight

One of the best outcomes of redirection as part of a No-Drama Discipline strategy is that it helps develop personal insight in our children. The reason is that instead of simply commanding and demanding that our kids meet our expectations, we ask them to notice and reflect on their own feelings and their responses to difficult situations. This can be difficult, as you know, since a child's upstairs brain is not only the last to develop, but it is often off-line in disciplinary moments. But with practice and insight-building conversations—like the ones we've been discussing and will explain at greater length in the next chapter—children can become more aware and understand themselves more fully. They can develop personal mindsight that allows them to better understand what they're feeling, and have more control over how they respond in difficult situations.

For young children we might facilitate this process simply by naming the emotions we observe: "When she took away the doll, it looked like you felt really mad. Is that right?" For older kids, open-ended questions are better, even if we have to "lead the witness" toward self-understanding: "I was watching you just before you blew up at your brother, and it looked like you were getting more and more annoyed that he was badgering you. Is that what you were feeling?" The hope is that his response is something like, "Yeah! And it makes me so mad when he . . ." Every time a child gets specific and discusses his own emotional experience, he gains more insight into himself and deepens his own self-understanding. That's a reflective conversation that cultivates mindsight. And such a focus on his insight can help him move toward the second desired outcome of redirection.

Outcome #2: Empathy

Along with developing insight into themselves, we want our kids to develop the other aspect of mindsight, empathy. The science of neuroplasticity teaches us that repeated practice of this reflection, as in our reflective dialogues with others, activates our mindsight circuitry. And with repeated focus of attention on our inner mental life, it also changes the wiring in the brain and builds and strengthens the empathic, other-centered part of the upstairs brain—what scientists call the social engagement circuitry of the prefrontal cortex. This is the part of the brain that makes mindsight maps not only of ourselves for insight and of others for empathy, but also of "we" for morality and mutual understanding. That's what mindsight circuits create. So we want to give kids lots of practice reflecting on how their actions impact others, seeing things from another's point of view, and developing awareness of others' feelings.

Simply asking questions and helping our children make observations like these will be much more effective than preaching sermons, delivering lectures, or giving consequences. The human brain is capable of extending itself in a way that allows us to comprehend the experiences of the people around us and even sense our connections as part of a "we" that develops with them. That's how we experience not only empathy, but the important sense of our interconnectedness, the integrated state that is the basis of moral imagination, thinking, and action.

So the more we give our kids practice at considering how someone else feels or experiences a situation, the more empathic and caring they will become. And as these circuits of insight and empathy develop, they naturally set the foundation for morality, our inner sense of being not only differentiated, but linked into a larger whole. That's integration.

Outcome #3: Integration and the Repair of Ruptures

After helping our kids to consider their own feelings and then reflect on how their actions impacted others, we want to ask them what they can do to create integration as they repair the situation and make things right. Which part of the brain do we appeal to now? You guessed it: the upstairs brain, with its responsibility for empathy, morality, considering the consequences of our decisions, and controlling emotions.

We appeal to the upstairs brain by asking questions, in this case about repairing a situation. "What can you do to make it right? What positive step can you take to help fix this? What do you think needs to happen now?" Repair builds on insight and empathy to then move to the mindsight map of "we" as a connection is reestablished with the other person. Once we've led our children toward empathy and insight, we want to aim for the outcome of taking action to address not only the situation their behavior has impacted, but also the other person and, ultimately, the relationship itself.

Taking action after hurting someone or making a bad decision isn't easy for any of us, including our kids. Especially when children are little, or if they have a particularly shy temperament, parents may need to support them and help them with their apology. Sometimes it's fine for the parent to actually deliver the apology *for* the child. You two can agree on the message beforehand. After all, not much good comes from forcing a child to offer an inauthentic apology when he's not yet ready, or forcing him to apologize when doing so is going to flood his nervous system with anxiety. It comes back to asking whether your child is ready. Sometimes we have to wait for a child to be in the right frame of mind.

It's never easy to go back and try to make up for a mistake. But No-Drama Discipline allows us to help kids learn to do so. It aims at achieving these three outcomes: focusing on giving our children practice at better understanding themselves with insight, seeing things from the perspective of others with empathy, then taking steps to improve a particular situation where they've done something wrong. When children deepen their ability to know themselves, consider the feelings of others, and take action toward repairing a situation, they build and strengthen connections within the frontal lobe, which allows them to better know themselves and get along with others as they move into adolescence and adulthood. Basically, you are teaching your child's brain how to make mindsight maps of "me," "you," and "we."

1-2-3 Discipline in Action

Life gives us opportunity after opportunity to build the brain. That's what we saw when Roger talked to his daughter about monopolizing her sister's playdate. He could've easily called out to his daughter something like, "Allie, why don't you give Katie and Gina some time to themselves?" In doing so, though, he would have missed an opportunity to teach Allie and help build her brain.

His response, instead, offered a 1-2-3 approach. By initiating a conversation with his daughter ("Do you see that Katie isn't happy?") rather than laying down the law, he focused on the one definition of discipline: teaching. He also worked from the two key principles. First, he made sure his daughter was *ready* by making her feel listened to without judgment ("I totally agree that it's not your fault"). And second, he *avoided being overly rigid* and even asked Allie for help in coming up with a good response to the situation. And he achieved the three outcomes, helping his daughter think about her own actions ("Why do you think she might be upset?"), her sister's feelings ("If she were standing here and told us how she felt, what would she say?"), and what response she could take to best respond as an integrative repair to the situation ("Let's come up with a plan").

The approach works with older children as well. Let's look at an example of how one couple applied it with their middle schooler.

At every major gift-giving occasion over the past year, Nila had consistently written the phrase "cell phone" at the top of her wish list. She repeatedly told her parents, Steve and Bela, that "all" the other kids had phones. Her mom and dad held out longer than many of their friends, but when she turned twelve, they relented. After all, Nila was reasonably responsible, she was spending more time independently from her parents, and a phone would make things more convenient for everyone. They took all the measures they knew were important—disabling the phone's Internet capabilities, downloading apps that would filter out dangerous content, talking with her about issues like privacy and security—then moved on to this next phase in their parenting lives.

During the first few months, Nila made her parents' decision look good. She kept track of the phone and used it appropriately, and they learned that they hadn't overestimated the convenience factor.

But one night Bela heard Nila coughing an hour after lights-out, so she opened the door to her room to check on her. The blue glow hovering over Nila's bed instantly disappeared, but it was obviously too late. She was busted.

Bela flipped on the overhead light, and before she could say anything, Nila hurried to explain: "Mom, I was worried about the test and couldn't sleep, so I was just trying to get my mind on something else."

Bela knew better than to overreact, especially when her primary goal in that moment was to get her daughter back to sleep, so she first connected: "I can understand needing to get your mind on something else. I hate it when I can't sleep." Then she simply said, "But let's talk about it tomorrow. Hand me the phone, and I want you to go right to sleep."

When Bela told Steve, she learned that he had had a similar interaction with Nila just the previous week when Bela was gone, and he'd forgotten to mention it. So now they had two cases of their daughter blatantly disregarding rules about cell phone usage and sleep.

Taking a 1-2-3 approach, Steve and Bela focused on the one definition of discipline. What lesson did they want to teach here? They wanted to emphasize the importance of honesty, responsibility, trust, and following the rules the family members have all agreed to. As they considered how to respond to Nila's infractions, they kept this definition front and center in their minds.

Then they focused on the two principles. Bela demonstrated the first one—making sure her daughter was ready—when she simply took Nila's phone and asked her to go to sleep. Late at night, when everyone is tired and a child is up later than she should be, is rarely the best time to teach a lesson. Lecturing Nila right then would have likely turned into all kinds of drama, leaving both mother and daughter frustrated and angry. Not exactly a recipe for going right to sleep, or teaching a lesson, either. The better strategy was to wait for the next day, when Bela and Steve could find the right moment to address the issue. Not during the morning rush to eat breakfast and make lunches, but after dinner when everyone could discuss the issue calmly and from a fresh perspective.

As for their specific response, this is where the second principle came in: be consistent, but not rigid. Consistency is of course crucial. Steve and Bela had taken a clear stance about the importance of Nila being honest and responsible with her phone, and at least in this instance, she hadn't lived up to their agreement. So they needed to address that lapse with a consistent response.

But in doing so, they didn't want to make a rigid, snap decision that overshot the mark. Their first reaction was to take the phone away altogether. But once they talked, and calmer heads prevailed, they recognized that in this case, that response would be too drastic. Outside of this one problem, Nila had acted responsibly with her phone. So rather than taking it away, they decided to discuss the issue with Nila, asking for her help coming up with policies to address the situation. In fact, she was the one who came up with a fix that was easy for everyone: she would leave her phone outside her room when she goes to bed. Then she wouldn't be tempted to check it every time it lit up—and Mom and Dad could be assured that Nila was recharging while her phone was as well.

This response made sense given Nila's general good decision making. They all agreed that if more problems arose, or if she demonstrated more extreme misuse of the phone, Steve and Bela would hold on to the phone except for certain prescribed times of the day.

With this response, which respected Nila enough to work *with* her and collaborate while still enforcing boundaries, Steve and Bela presented a consistent, united front that adhered to their rules and expectations, without becoming rigid and disciplining in ways that wouldn't benefit their daughter, the situation, or their relationship with her.

As a result, they gave themselves a much better chance at achieving their three desired outcomes: insight, empathy, and integrated repair. They helped encourage insight in their daughter by the collaborative approach they took in asking questions and engaging in dialogue. The questions focused on helping Nila pause and think about her decision to power up her phone when she wasn't supposed to: "How do you feel inside when you're doing something you know you shouldn't? Or when we walk in and see you on your phone? What do you think we feel about it?" Other questions led to insight into better options in the future: "*The next time you're having trouble sleeping, what could you do instead of being on your phone?*" With questions like these, Nila's parents helped increase her personal insight and build her upstairs brain, allowing her to develop an internal compass and become more insightful in the future. Plus, by approaching the issue in a way that respected her and her desires, they increased the chances that Nila will come and talk with them about even bigger issues later, as she enters her teen years.

The empathy outcome in this situation is different from certain other discipline moments. Often when we encourage empathy in our children when they've made a bad decision, we try to lead them to think about the feelings of someone else who was hurt because of their behavior. In this case, no one was really harmed except for Nila herself, who lost some sleep. But Steve and Bela tried to lead her to understand that their trust in her had been dented, at least a bit. They knew better than to overdramatize the issue, or stoop to using guilt trips or self-pity, and they explicitly communicated to her that they weren't going to resort to these tactics. But they talked with her about how much their relationship with her means, and explained that it doesn't feel good when broken trust harms that relationship in any way.

This part of the discussion about the relationship is a focus on integration, the connection of different parts. Integration is what makes the whole greater than the sum of its parts, and it's what creates love in a relationship. Focusing on insight and empathy and then on their relationship thus led naturally to the third desired integrative outcome, repair. Once a breach in a relationship has been created, no matter how small, we want to repair it as soon as possible. Nila's parents needed to give her that chance. In their discussion about what policies to put in place about late-night cell phone use, they asked questions that helped her think about the relational effects of not following through on commitments. Again, they avoided manipulating her emotionally by making her feel guilty, and instead asked good-faith questions like "What's something you could do to help us feel good about the trust we have in you?" They had to "lead the witness" a bit, helping Nila think about trust-building actions she could take—like using her phone to just call and check in with her parents from time to time, or leaving it outside her room at night without having to be asked. In doing so she thought about ways she could be intentional in rebuilding her parents' trust in her.

Notice that this issue with Nila falls into the category of typical behaviors that parents have to deal with on a daily basis. At times, there are behavioral challenges where it can be helpful to involve professionals. More extreme behaviors that are difficult to handle and that last for longer periods of time can sometimes be a sign that something else is going on. If your child frequently experiences intense emotional reactivity that does not respond to repair efforts, it can be helpful to talk with a pediatric psychotherapist or child development specialist who can supportively explore the situation with you to see whether you and your child could benefit from some intervention. In our experience, children who display frequent and intense reactivity may be struggling with more innate challenges related to sensory integration, attention and/or impulsivity, or mood disorders. Additionally, a history of trauma, a really difficult experience from the past, or relational mismatches between parent and child can play a role in behavioral struggles, as they reveal an underlying challenge with self-regulation that may at times be a source of repeated ruptures in a relationship. We would encourage you to seek the help of someone who can help you walk through these questions and guide you and your child on the path toward optimal development.

In most discipline situations with your child, though, simply taking a Whole-Brain approach will lead to more cooperation from your child and more peace and serenity in your household. 1-2-3 discipline isn't a formula or a set of rules to be strictly followed. You don't have to memorize it and inflexibly follow it. We're simply giving you guidelines to keep in mind when it comes time for redirection. By reminding yourself about the definition and purpose of discipline, the principles that should guide it, and your desired outcomes, you'll give yourself a much better chance of disciplining your kids, of teaching them, in a way that leads to more cooperation from them and better relationships among all members of the family.

CHAPTER 6

Addressing Behavior:
As Simple as R-E-D-I-R-E-C-T

nna's eleven-year-old, Paolo, called her from school and asked whether he could go home with his friend Harrison that afternoon. The plan, Paolo explained, was to walk to Harrison's, where the boys would do homework, then play until dinner. When Anna asked whether Harrison's parents were aware of the plan, Paolo assured her they were, so Anna told him she'd pick him up before dinner.

However, when Anna texted Harrison's mother later that afternoon, telling her she'd be picking up Paolo in a few minutes, Harrison's mother revealed that she was at work. Anna then learned that Harrison's father hadn't been home, either, and that neither of them knew of the boys' plan for Paolo to come over.

Anna was mad. She knew there might have been some sort of miscommunication, but it really looked to her like Paolo had been dishonest. At best he had misunderstood the plan, in which case he should have let her know when he realized that Harrison's parents wouldn't be home and hadn't been contacted. At worst he had outright lied to her.

Once she and Paolo were in the car on the way home from Harrison's, she felt like launching into him, leveling consequences and angrily lecturing him about trust and responsibility.

But that's not what she did.

Instead, she took a Whole-Brain approach. Since her son was older and he wasn't in a reactive state of mind, the "connect" part of her approach simply entailed hugging him and asking whether he'd had a good time. Then she showed him the respect of communicating with him directly. She told him about her text with Harrison's mother, then said simply, "I'm glad you and Harrison have so much fun together. But I have a question. I know you know how important trust is in our family, so I'm wondering what happened here." She spoke in a calm tone, one that didn't communicate harshness and instead expressed her lack of understanding and her curiosity about the situation.

This curiosity-based approach, where she began by giving her son the benefit of the doubt, helped Anna decrease the drama from the discipline situation. Even though she was angry, she avoided immediately jumping to the conclusion that the boys had purposely deceived their parents. As a result, Paolo could hear his mother's question without feeling directly accused. Plus, her curiosity put the responsibility of accounting for himself squarely on Paolo's own shoulders, so he had to think about his decision making, which gave his upstairs brain a little bit of exercise. Anna's approach showed Paolo that she worked from the assumption that he would make good decisions most of the time, and that she was confused and surprised when it appeared that he hadn't.

In this case, by the way, he hadn't made good decisions. He explained to his mother that Harrison had thought his father would be home, but when the boys arrived, Harrison's father wasn't there. He acknowledged that he should have let her know right away, but he just hadn't. "I know, Mom. I should've told you nobody else was home. Sorry."

Then Anna could respond and move from connection to redirection, saying something like, "Yes, I'm glad you're clear that you should have told me. Tell me more about why that didn't happen." But she knew she wanted her redirection to be about more than just addressing this one behavior. She rightly recognized this moment as another opportunity to build important personal and relational skills in her son, and to help him understand that his actions had made a little dent in her trust and deviated from their family agreement to always check in if plans change. That's why, before she turned to redirection, she checked herself.

Before You Redirect: Keep Calm and Connect

Have you seen that British poster from World War II that's become so popular? The one that says, "Keep Calm and Carry On"? That's not a bad mantra to have at the ready when your child goes ballistic—or before *you* do. Anna recognized the importance of keeping calm when she addressed her problem with Paolo's behavior. Blowing up and yelling at her son wouldn't have done anyone any good. In fact, it would have alienated Paolo and become a distraction from what was important here: using this disciplinary moment to address his behavior, and to teach.

We'll discuss many redirection strategies below, looking at different ways to redirect children when they've made bad decisions or completely lost control of themselves. But before you decide on which redirection strategies to use as you redirect your kids toward using their upstairs brains, you should first do one thing: check *yourself*. Remember, just as it's important to ask, "Is my child ready?" it's also essential that you ask, "Am I ready?"

Imagine that you walk into your recently cleaned kitchen and find your four-year-old perched on the counter, an empty egg carton and a dozen broken shells by her side, stirring a sand bucket full of eggs. With her sand shovel! Or your twelve-year-old informs you, at 6:00 p.m. on Sunday, that his 3-D model of a cell is due the following morning. This despite the fact that he assured you that all his homework was done, then spent the afternoon playing basketball and video games with a friend.

In the middle of frustrating moments like these, the best thing you can do is to pause. Otherwise your reactive state of mind might lead you to begin yelling, or at least lecturing about the fact that a four-year-old (or twelve-year-old) ought to know better.

Instead, pause. Just pause. Allow yourself to take a breath. Avoid reacting, issuing consequences, or even lecturing in the heat of the moment.

We know it's not easy, but remember: when your kids have messed up in some way, you want to redirect them back toward their upstairs brain. So it's important to be in yours, too. When your three-year-old is throwing a tantrum, remember that she's only a small child with a limited capacity to control her own emotions and body. Your job is to be the adult in the relationship and carry on as the parent, as a safe, calm haven in the emotional storm. *How you respond to your child's behavior will greatly impact how the whole scene unfolds.* So before you redirect, check yourself and do your best to keep calm. That's a pause that comes from the upstairs brain but also reinforces the strength of your upstairs brain. Plus, when you show abilities like this to your children, they're more likely to learn such skills themselves.

Staying clear and calm during a pause is your first step.

Then remember to connect. It really is possible to be calm, loving, and nurturing while disciplining your child. And it's *so effective.* Don't underestimate how powerful a kind tone of voice can be as you initiate a conversation about the behavior you're wanting to change. Remember that, ultimately, you're trying to remain firm and consistent in your discipline while still interacting with your child in a way that communicates warmth, love, respect, and compassion. These two aspects of parenting can and should coexist. That was the balance Anna tried to strike as she spoke with Paolo.

As you've heard us affirm throughout the book, kids need boundaries, even when they're upset. But we can hold the line while providing lots of empathy and validation of the desires and feelings behind our child's behavior. You might say, "I know you really want another ice pop, but I'm not going to change my mind. It's OK to cry and be

sad and disappointed, though. And I'll be right here to comfort you while you're sad."

And remember not to dismiss a child's feelings. Instead, acknowledge the internal, subjective experience. When a child reacts strongly to a situation, especially when the reaction seems unwarranted and even ridiculous, the temptation for the parent is to say something like "You're just tired" or "It's not that big of a deal" or "Why are you so upset about this?" But statements like these minimize the child's experience—her thoughts, feelings, and desires. It's much more emotionally responsive *and effective* to listen, empathize, and really understand your child's experience before you respond. Your child's desire might seem absurd to you, but don't forget that it's very real to him, and you don't want to disregard something that's important to him.

So when it's time to discipline, keep calm and connect. Then you can turn to your redirection strategies.

Strategies to Help You R-E-D-I-R-E-C-T

For the remainder of this chapter we'll focus on what you may have been waiting for: specific, No-Drama redirection strategies you can take once you've connected with your children and want to redirect them back to their upstairs brain. To help organize the strategies, we've listed them as an acronym:

Reduce words
Embrace emotions
Describe, don't preach
Involve your child in the discipline
Reframe a no into a conditional yes
Emphasize the positive
Creatively approach the situation
Teach mindsight tools

Before we get into specifics, let us be clear: this isn't a list you need to memorize. These are simply categorized recommendations that the parents we've worked with over the years have found to be the most helpful. (We've included the list, by the way, in the Refrigerator Sheet at the back of the book.) As always, you should keep all of these various strategies as different approaches in your parental tool kit, picking and choosing the ones that make sense in various circumstances according to the temperament, age, and stage of your child, as well as your own parenting philosophy.

Redirection Strategy #1: Reduce Words

In disciplinary interactions, parents often feel the need to point out what their kids did wrong and highlight what needs to change next time. The kids, on the other hand, usually already know what they've done wrong, especially as they get older. The last thing they want (or, usually, need) is a long lecture about their mistakes.

WHAT A PARENT SAYS:

WHAT THE CHILD HEARS:

We strongly suggest that when you redirect, you resist the urge to overtalk. *Of course* it's important to address the issue and teach the lesson. But in doing so, keep it succinct. Regardless of the age of your children, long lectures aren't likely to make them want to listen to you more. Instead, you'll just be flooding them with more information and sensory input. As a result, they'll often simply tune you out.

With younger children, who may *not* have learned yet what's OK and what's not, it's even more important that we reduce our words. They often just don't have the capacity to take in a long lecture. So instead, we need to reduce our words.

If your toddler, for instance, hits you because she's angry that she doesn't have your attention while you're attending to your other child, there's simply no reason to go off on a long, drawn-out oration about why hitting is a bad response to negative emotions. Instead, try this four-step approach that addresses the issue and then moves on, all without using more than a few words:

ADDRESSING TODDLER MISBEHAVIOR IN FOUR STEPS

STEP 1: Connect and address the feelings behind the behavior

Step 2: Address the behavior

Step 3: Give alternatives

Step 4: Move on

By addressing the child's actions and then immediately moving on, we avoid giving too much attention to the negative behavior and instead quickly get back on the right track.

For younger and older kids both, avoid the temptation to talk too much when you discipline. If you do need to cover an issue more fully, try to do so by asking questions and then listening. As we'll explain below, a collaborative discussion can lead to all kinds of important teaching and learning, and parents can accomplish their disciplinary goals without talking nearly as much as they typically do.

The basic idea here is akin to the concept of "saving your voice."

Politicians, businesspeople, community leaders, and anyone else who depends on effective communication to achieve their goals will tell you that often there are times when they strategically save their voice, holding back on how much they say. They don't mean their literal voice, as if they'll make their throats hoarse by talking so much. They mean they try to resist addressing the small points in a discussion or a voting meeting, so that their words will matter more when they want to address the really important issues.

It's the same with our kids. If they hear us incessantly telling them what to do and what not to do, and then once we've made our point we keep making it over and over again, they will sooner or later (and probably sooner) stop listening. If, on the other hand, we save our voice and address what we really care about, then stop talking, the words we use will carry much greater weight.

Want your kids to listen to you better? Be brief. Once you address the behavior and the feelings behind the behavior, move on.

Redirection Strategy #2: Embrace Emotions

One of the best ways to address misbehavior is to help kids distinguish between their feelings and their actions. This strategy is related to the concept of connection, but we're actually making a completely different point here.

When we say to embrace emotions, we mean that during redirection, parents need to help their kids understand that their feelings are neither good nor bad, neither valid nor invalid. They simply *are*. There's nothing wrong with getting angry, being sad, or feeling so frustrated that you want to destroy something. But saying it's OK to *feel* like destroying something doesn't mean it's OK to actually do it. In other words, it's what we *do* as a result of our emotions that determines whether our behavior is OK or not OK.

So our message to our children should be, "You can feel whatever you feel, but you can't always do whatever you want to do." Another way to think about it is that *we want to say yes to our kids' desires, even when we need to say no to their behavior and redirect them toward appropriate action.*

So we might say, "I know you want to take the shopping cart home. That would be really fun to play with. But it needs to stay here at the store so other shoppers can use it when they come." Or we might say, "I totally get it that you feel like you hate your brother right now. I used to feel that way about my sister when I was a kid and was really mad at her. But yelling 'I'm going to kill you!' isn't how we talk to each other. It's perfectly fine to be mad, and you have every right to tell your brother about it. But let's talk about other ways to express it." Say yes to the feelings, even as you say no to the behavior.

When we don't acknowledge and validate our kids' feelings, or when we imply that their emotions should be turned off or are "no big deal" or "silly," we communicate the message, "I'm not interested in your feelings, and you should not share them with me. You just stuff those feelings right on down." Imagine how that impacts the relationship. Over time, our children will stop sharing their internal experiences with us! As a result, their overall emotional life will begin to constrict, leaving them less able to fully participate in meaningful relationships and interactions.

Even more problematic is that a child whose parents minimize or deny her feelings can begin to develop what can be called an "incoherent core self." When she experiences intense sadness and frustration, but her mother responds with statements like "Relax" or "You're fine," the child will realize, if only at an unconscious level, that her internal response to a situation doesn't match the external response from the person she trusts most. As parents, we want to offer what's called a "contingent response," which means that we attune our response to what our child is actually feeling, in a way that validates what's happening in her mind. If a child experiences an event and the response from her caregiver is consistent with it—if it's a match—then her internal experience will make sense to her, and she can understand herself, confidently name the internal experience, and

INSTEAD OF SQUELCHING EMOTIONS...

SAY YES TO THE FEELINGS AND NO TO THE BEHAVIOR

communicate it to others. She'll be developing and working from a "coherent core self."

But what happens if that match isn't there and her mother's response is inconsistent with the daughter's experience of the moment? One mismatch isn't going to have long-lasting effects. But if over and over again when she gets upset she is told something like "Stop crying" or "Why are you so upset? Everyone else is having fun," she's going to begin to doubt her ability to accurately observe and comprehend what's going on inside her. Her core self will be much more incoherent, leaving her confused, full of self-doubt, and disconnected from her emotions. As she grows into an adult, she may often feel that her very emotions are unjustified. She might doubt her subjective experience, and even have a hard time knowing what she wants or feels at times. So it really is crucial that we embrace our children's emotions and offer a contingent response when they are upset or out of control.

One bonus to acknowledging our children's feelings during redirection is that doing so can help kids more easily learn whatever lesson we're wanting to teach. When we validate their emotions and acknowledge the way *they* are experiencing something—really seeing it through their eyes—that validation begins to calm and regulate their nervous system's reactivity. And when they are in a regulated place, they have the capacity to handle themselves well, listen to us, and make good decisions. On the other hand, when we deny our kids' feelings, minimize them, or try to distract our kids from them, we prime them to be easily dysregulated again, and to feel disconnected from us, which means they'll operate in a heightened state of agitation and be much more likely to fall apart, or shut down emotionally, when things don't go their way.

What's more, if we're saying no to their emotions, kids aren't going to feel heard and respected. We want them to know that we're here for them, that we'll always listen to how they feel, and that they can come to us to discuss anything they're worried about or dealing with. We don't want to communicate that we're here for them only when they're happy or feeling positive emotions.

So in a disciplinary interaction, we embrace our kids' emotions, and we teach them to do the same. *We want them to believe at a deep level that even as we teach them about right and wrong behavior, their feelings and experiences will always be validated and honored. When kids feel this from their parents even during redirection, they'll be much more apt to learn the lessons the parents are teaching, meaning that over time, the overall number of disciplinary moments will decrease.*

Redirection Strategy #3: Describe, Don't Preach

The natural tendency for many parents is to criticize and preach when our kids do something we don't like. In most disciplinary situations, though, those responses simply aren't necessary. Instead, we can simply describe what we're seeing, and our kids will get what we're saying just as clearly as they do when we yell and disparage and nitpick. And they'll receive that message with much less defensiveness and drama.

With a toddler we might say something like, "Uh-oh, you're throwing the cards. That makes it hard to play the game." To an older child we can say, "I still see dishes on the table," or "Those sound like some pretty mean words you're using with your brother." Simply by stating what we observe, we initiate a dialogue with our children that opens the door to cooperation and teaching much better than an immediate reprimand like "Stop talking to your brother that way."

The reason is that even young children know wrong from right in most situations. You've already taught them what's acceptable behavior and what's not. Often, then, all you need to do is call attention to the behavior you've observed. This is essentially what Anna did when she said to Paolo, "I know you know how important trust is in our family, so I'm wondering what happened here." *Kids don't need their parents to tell them not to make bad decisions. What they need is for their parents to redirect them, helping them recognize the bad decisions they're making and what leads up to those decisions, so they can correct themselves and change whatever needs to be changed.*

For all kids, and especially younger children and toddlers, you are of course teaching them good from bad, right from wrong. But again, a short, clear, direct message is going to be much more effective than a longer, overexplained one. And even with young children, a simple statement of observation will typically get your point across—and invite a response from them, either verbally or behaviorally.

The idea here isn't that a description of what you see will be some sort of magical phrase that stops bad behavior in its tracks. We're simply saying that parents should, as we put it in Chapter 5, "think about the how" and be intentional about *how* they say what needs to be said.

It's not that the phrase "Looks like Johnny wants a turn on the swing" is communicating something fundamentally different from the phrase "You need to share." But the former offers several distinct advantages over the latter. First, it avoids putting a child on the defensive. She might still feel the need to defend herself, but not to the same degree as if we were to reprimand her or tell her what she's doing wrong.

Second, describing what we see puts the onus for deciding how to respond to the observation on the child, thus exercising his upstairs brain. That's how we help him develop an internal compass, a skill that can last a lifetime. When we say, "Jake is feeling left out; you need to include him," we are definitely getting our message across. But we're doing all the work for our child, not allowing him to increase his inner skills of problem solving and empathy. If instead we simply say, "Look at Jake sitting over there while you and Leo play," we give our child the opportunity to consider the situation for himself, and determine what needs to happen.

Third, describing what we see initiates a conversation, thereby implying that when our child does something we don't like, our default response will be to visit with her about it, allow her to explain, and gain some insight. Then we can give her a chance to defend herself or apologize if necessary, and to come up with a solution to whatever problem her behavior might have caused.

INSTEAD OF COMMANDING AND DEMANDING...

DESCRIBE WHAT YOU SEE

INSTEAD OF CRITICIZING AND ATTACKING...

DESCRIBE WHAT YOU ARE SEEING

"What's going on?" "Can you help me understand?" "I can't figure this out." These can be powerful phrases when we're teaching our kids. When we point out what we see, then ask our kids to help us understand, it opens up the opportunity for cooperation, dialogue, and growth.

Do you see how the two responses, even though their *content* isn't all that different, would be apt to garner very different responses from the children, simply because of *how* the parents communicated their message? Once the parents describe what they've observed and ask for help in understanding, they can pause and allow the child's brain to do its work. Then they can take an active role in their response.

This redirection strategy leads directly into the next one, which is all about making discipline a collaborative, mutual process, rather than a top-down imposition of parental will.

Redirection Strategy #4: Involve Your Child in the Discipline

When it comes to communicating in a disciplinary moment, parents have traditionally done the talking (read: lecturing), and children have done the listening (read: ignoring). Parents have typically worked from an unexamined assumption that this one-directional, monologue-based approach is the best—and only viable—option to consider.

PARENT → CHILD

Many parents these days, however, are learning that discipline will be much more respectful—and, yes, effective—if they initiate a collaborative, reciprocal, bidirectional *dialogue*, rather than delivering a monologue.

PARENT ↔ CHILD

We're not saying that parents should forgo their roles as authority figures in the relationship. If you've read this far in the book, you know that we definitely don't advocate that. But we do know that *when children are involved in the process of discipline, they feel more respected, they buy into what the parents are promoting, and they are therefore more apt to cooperate and even help come up with solutions to the problems that created the need for discipline in the first place. As a result, parents and children work as a team to figure out how best to address disciplinary situations.*

Remember our discussion of mindsight, and the importance of helping kids develop insight into their own actions and empathy for others? Once you've connected and your child is ready and receptive, you can simply initiate a dialogue that leads first toward insight ("I know you know the rule, so I'm wondering what was going on for you that led you to this") and then toward empathy and integrative repair ("What do you think that was like for her, and how could you make things right?").

For example, let's say your eight-year-old becomes out-of-control furious because his sister is going on *another* playdate, and he feels like he "never gets to do *anything*!" In his anger, he throws your favorite sunglasses across the room and breaks them.

Once you've calmed down and connected with your son, how do you want to talk with him about his actions? The traditional approach is to offer a monologue where you say something like, "It's OK to get mad—everyone does—but when you're angry you still need to control your body. We don't break other people's things. The next time you're that mad, you need to find an appropriate way to express your big feelings."

Is there anything wrong with this communication style? No, not at all. In fact, it's full of compassion and a healthy respect for your child and his emotions. But do you see how it's based on top-down, one-directional communication? You are imparting the important information, and your child is receiving it.

What if, instead, you involved him in a collaborative dialogue that asked him to consider how best to address the situation? Maybe you would begin with the D from R-E-D-I-R-E-C-T and merely *describe* what you saw, then ask him to respond: "You got so mad a while ago. You grabbed my glasses and threw them. What was going on?"

Since you will have already connected, listened, and responded to his feelings about his sister's playdate, he can now focus on your question. Most likely he'll come back to his anger and say something like, "I was just so mad!"

Then you can simply describe, being intentional with your tone (since the how matters), what you saw: "Then you threw my glasses." Here's where you're likely to get some sort of "Sorry, Mom."

At this point you can move to the next phase of the conversation and focus explicitly on teaching: "We all get mad. There's nothing wrong with getting angry. But what could you do the next time you're that mad?" Maybe you could even smile and throw in some subtle humor he'd appreciate: "You know, besides destroying something?" And the conversation could go on from there, with you asking questions that help your young son think about issues like empathy, mutual respect, ethics, and handling big emotions.

Notice that the overall message remains the same, whether you offer a monologue or initiate a dialogue. But when you involve your child in the discipline, you give him the opportunity to think about his own actions, and whatever resulted from them, at a much deeper level.

You help him recruit more complex neural pathways that build mindsight capacities, and the result is deeper and longer-lasting learning.

Involving your kids in the discipline discussion is also a great way to dial back any patterns or behaviors that may have unintentionally been set up in your home. A one-directional, top-down discipline

INSTEAD OF DELIVERING A MONOLOGUE...

INVOLVE YOUR CHILD IN THE DISCIPLINE

approach might lead you to storm into the living room and declare, "You're spending way too much time on video games these days! From now on, no more than fifteen minutes a day." You can imagine the response you might receive.

What if, instead, you waited until dinnertime, and once everyone was at the table, you said, "I know you've been getting to play video games a lot lately, but that's not really working very well. It puts off homework, and I also want to make sure you're spending time on other activities as well. So we need to come up with a new plan. Any ideas?"

You will probably still experience resistance when you broach the possibility of curtailing screen time. But you will have initiated a discussion about the issue, and when your kids know that you're talking about cutting back, they'll definitely be invested in being a part of the conversation to determine what limits will be set. You can remind them that you will be making the final decision, but let them see that you're inviting their input because you respect them, want to consider their feelings and desires, and believe they are helpful problem solvers. Then, even if they don't love the final call you make, they'll know they were at least considered.

The same would go for any number of other issues: "I know we've been doing homework after dinner, but that's not been working well, so we need a new plan. Any ideas?" Or "I've noticed that you're not too happy about having to practice piano before school in the mornings. Is there a different time when you'd feel better about practicing? What would work for you?" Often they'll come up with the same solution you would have imposed on the situation anyway. But they will have exercised their upstairs brain to do so and felt your respect along the way.

One of the best results from involving kids in the discipline process is that frequently they'll come up with great new ideas for solving a problem, ideas you hadn't even considered. Plus, you might be shocked to find out how much they are willing to bend to bring about a peaceful resolution to a standoff.

INSTEAD OF COMMANDING AND DEMANDING...

INVOLVE YOUR KIDS IN THE DISCIPLINE

Tina tells the story of a time when her four-year-old absolutely *had* to have a treat—specifically, a bag of fruit snacks—at nine-thirty in the morning. She told him, "Those fruit snacks are delicious, aren't they? You can have them after you have a good lunch in a little while."

He didn't like Tina's plan and began to cry and complain and argue. She responded by saying, "It's really hard to wait, isn't it? You want the fruit snacks, and I want you to have a healthy lunch first. Hmmm. Do you have any ideas?"

She saw his little cognitive wheels turn for a few seconds, then his eyes got big with excitement. He called out, "I know! I can have one *now* and save the rest for after lunch!"

He felt empowered, the power struggle was averted, and Tina was able to give him an opportunity to solve a problem. And all it cost her was allowing him to have *one* fruit snack. Not such a big deal.

Again, there are of course times that you can't give any wiggle room, and there may be times to allow your child to deal with a no or give him the opportunity to learn about waiting or handling disappointment. But usually when we involve the child in the discipline, it results in a win-win solution.

Even with very young children, we want to involve them as much as possible, asking them to reflect on their actions and consider how to avoid problems in the future: "Remember yesterday, when you got angry? You're not usually someone who hits and kicks. What happened?" With questions like these, you give your child the opportunity to practice reflecting on her behavior and developing self-insight. Granted, you may not get great answers from a young child, but you're laying the groundwork. The point is to let her think about her own actions.

Then you can ask her what she can do differently the next time she gets so mad. Discuss what she would like you to do to help her calm down. This type of conversation will deepen her understanding of the importance of regulating emotions, honoring relationships, planning ahead, expressing herself appropriately, and on and on. It will also communicate how important her input and ideas are to you.

She'll understand more and more that she's an individual, separate from you, and that you are interested in her thoughts and feelings. Every time you involve your children in the process of discipline, you strengthen the parent-child bond, while also increasing the odds that they'll handle themselves better in the future.

Redirection Strategy #5: Reframe a No into a Conditional Yes

When you have to decline a request, it matters, once again, *how* you say no. An out-and-out no can be much harder to accept than a yes with conditions. No, especially if said in a harsh and dismissive tone, can automatically activate a reactive state in a child (or anyone). In the brain, reactivity can involve the impulse to fight, flee, freeze, or, in extreme cases, faint. In contrast, a supportive yes statement, even when not permitting a behavior, turns on the social engagement circuitry, making the brain receptive to what's happening, making learning more likely, and promoting connections with others.

This strategy will differ according to the age of your children. To a toddler who is asking for more time at her grandmother's when it's time to leave, you can say, "Of course you can have more time with Nana. We need to go now, but Nana, would it be OK if we came back to your house this weekend?" The child may still have trouble accepting no, but you're helping her see that even though she's not getting exactly what she wants right now, she'll be told yes again before too long. The key is that you've identified and empathized with a feeling (the desire to be with Nana) while creating structure and skill (acknowledging the need to leave now and delaying the gratification of the desire).

Or if your son can't get enough of the Thomas the Tank Engine hands-on display at the local toy store and is unwilling to set down Percy the Engine so you can exit the store, you can offer him a conditional yes. Try something like, "I know! Let's take Percy up to the saleswoman over there, and explain to her that you want her to hold him for you and keep him safe until we come back for story time on

INSTEAD OF AN OUTRIGHT NO...

REFRAME THE NO INTO A CONDITIONAL YES

Tuesday." The saleswoman will likely play along, and the whole potential fiasco can be avoided. What's more, you'll be teaching your child to develop a prospective mind, to sense the possibilities for the future and to imagine how to create future actions to meet present needs. These are executive functions that, when learned, can be skills that last a lifetime. You are offering guidance to literally grow the important prefrontal circuits of emotional and social intelligence.

Notice that this isn't at all about protecting kids from being frustrated or providing them with everything they want. On the contrary, it's about giving them practice at tolerating their disappointment when things inevitably don't go their way. They aren't attaining their desires in that moment, and you're assisting them as they manage their disappointment. You're helping them develop the resilience that will aid them every time they are told no throughout their lives. You're expanding their window of tolerance for not getting their way and giving them practice at delaying gratification. These are all prefrontal functions that develop in your child as you parent with the brain in mind. Instead of discipline simply leading to a feeling of being shut down, now your child will know, from actual experiences with you, that the limits you set often lead toward learning skills and imagining future possibilities, not imprisonment and dismissal.

The strategy is effective for older children (and even adults) as well. None of us like to be simply told no when we want something, and depending on what else has been happening, a no may even push us over the edge. So instead of offering an outright refusal, we can say something like, "There's a lot happening today and tomorrow, so yes, let's invite your friend over, but let's do it on Friday, when you'll have more time with him." That's a lot easier to accept, and it gives a child practice in handling the disappointment, as well as in delaying gratification.

Say, for instance, a group of your nine-year-old's friends are going to a concert to see the latest pop sensation, who, in your opinion, represents all the things you want your daughter *not* to emulate. Regardless of how you deliver the news, she's not going to be happy to hear that she's not going to the concert. But you can at least mitigate some of the drama by being proactive and getting ahead of the curve on the issue.

You might, for example, ask her about upcoming concerts she'd like to attend, and offer to take her and a friend to the movies in the meantime. If you want to go the extra mile, you could even get online and look for a different concert she'd be interested in attending in the near future. Pay close attention to your tone of voice. Particularly if you're having to deny a child something she really wants, it's important that you avoid coming across as patronizing or overly dogmatic in your opinion. Again, we're not saying this strategy will make everything easy and keep your child from feeling angry, hurt, and misunderstood. But by coming up with some sort of conditional yes, rather than a simple "No, you're not going," you at least decrease the reactivity and show your child that you're paying attention to her desires.

Granted, there are times we simply have to deliver the dreaded outright no. But it's more often the case that we can find ways to avoid having to turn our kids down without at least finding some measure of a yes that we can also deliver. After all, the things kids want are often the things we want for them, too—just at a different time. They may want to read more stories, or play with their friends, or eat ice cream, or play on the computer. These are all activities we want them to enjoy at some point as well, so usually we can easily find an alternative time to make it happen.

In fact, there's an important place for negotiation in parent-child interactions. This becomes more and more important as kids get older. When your ten-year-old wants to stay up a little later and you've said no, but then he points out that tomorrow is Saturday and he promises to sleep an hour later than usual, that's a good time to at least rethink your position. Obviously, there are some non-negotiables: "Sorry, but you can't put your baby sister in the dryer, even if you do line it with pillows." But compromise isn't a sign of weakness; it's evidence of respect for your child and his desires. In addition, it gives him an opportunity for some pretty complex thinking, equipping him with important skills about considering not only what he wants, but also what others want, and then making good arguments based on that information. And it's a *lot* more effective in the long run than just saying no without considering other alternatives.

Redirection Strategy #6: Emphasize the Positive

Parents often forget that discipline doesn't always have to be negative. Yes, it's usually the case that we're disciplining because something less than optimal has occurred; there's a lesson that needs to be learned or a skill that needs to be developed. But one of the best ways to deal with misbehavior is to focus on the positive aspects of what your kids are doing.

For example, think about that bane of parental existence, whining. Who doesn't get tired of hearing our kids shift to that droning, complaining, singsong tone of voice that makes us grit our teeth and want to cover our ears? Parents often respond by saying something like, "Stop whining!" Or maybe they'll get creative and say, "Turn down the whine," or "What's that? I don't speak whine. You'll have to tell me in another language."

We're not saying these are the worst possible approaches. It's a problem, though, when we resort to negative responses, because it gives all of our attention to the behavior we don't want to see repeated.

Instead, what if we emphasized the positive? Instead of "No whining," we could say something like, "I like it when you talk in your normal voice. Can you say that again?" Or be even more direct in teaching about effective communication: "Ask me again in your powerful, big-boy voice."

The same idea goes for other disciplinary situations. Instead of focusing on what you *don't* want ("Stop messing around and get ready, you're going to be late for school!"), emphasize what you *do*

INSTEAD OF FOCUSING ON THE PROBLEM...

EMPHASIZE THE POSITIVE

want ("I need you to brush your teeth and find your backpack"). Rather than highlighting the negative behavior ("No bike ride until you try your green beans"), focus on the positive ("Have a few bites of the green beans, and we'll hop on the bikes").

There are plenty of other ways to emphasize the positive when you discipline. You may have heard the old suggestion to "catch" your kids behaving well and making good decisions. Anytime you see your older child, who's usually so critical of her younger sister, giving her a compliment, point it out: "I love it when you're encouraging like that." Or if your sixth grader has had a hard time getting his homework in on time, and you notice that he's making a special effort to work ahead on the report that's due next week, affirm him: "You're really working hard, aren't you? Thanks for thinking ahead." Or when your kids are laughing together rather than fighting, make a point of it: "You two are really having fun. I know you argue, too, but it's great how much you enjoy each other."

EMPHASIZE THE POSITIVE BY CATCHING YOUR KIDS BEHAVING WELL

In emphasizing the positive, you give your focus and attention to the behaviors you want to see repeated. It's a gentle way to also encourage those behaviors in the future without the interaction becoming about rewards or praise. Simply giving your attention to your child and stating what you see can be a positive experience unto itself.

We're not saying you're not going to have to address negative behaviors as well. Of course you are. But as much as possible, focus on the positive and allow your kids to understand, *and to feel from you*, that you notice and appreciate when they're making good decisions and handling themselves well.

Redirection Strategy #7: Creatively Approach the Situation

One of the best tools to keep ready in your parenting toolbox is creativity. As we've said time and again throughout the book, there's no one-size-fits-all discipline technique to use in every situation. Instead, we've got to be willing and able to think on our feet and come up with different ways to handle whatever issue arises. As we put it in Chapter 5, parents need response flexibility, which allows us to pause and consider various responses to a situation, applying different approaches based on our own parenting style and each individual child's temperament and needs.

When we exercise response flexibility, we use our prefrontal cortex, which is central to our upstairs brain and the skills of executive functions. Engaging this part of our brain during a disciplinary moment makes it far more likely that we'll also be able to conjure up empathy, attuned communication, and even the ability to calm our own reactivity. If, on the other hand, we become *inflexible* and remain on the rigid bank of the river, we become much more reactive as parents and don't handle ourselves as well. Ever had that kind of moment? We have, too. Our downstairs brain will take charge and run the show, allowing our reactive brain circuitry to take over. That's why it's so important that we strive for response flexibility and crea-

tivity, especially when our kids are out of control or making bad decisions. Then we can come up with creative and innovative ways to
approach difficult situations.

For example, humor is a powerful tool when a child is upset. Especially with younger children, you can completely change the dynamics of an interaction simply by talking in a silly voice, falling
down comically, or using some other form of slapstick. If you are six
years old and furious with your father, it's not as easy to stay mad at
him if he's just tripped over a toy in the living room and enacted the
longest, most drawn-out fall to the ground you've ever seen. Likewise, leaving the park is a lot more fun if you get to chase Mom to the
car while she cackles and screams in pretend fear. Being playful is a
great way to break through a child's bubble of high emotion, so you
can then help him gain control of himself.

It applies to interactions with older kids, too; you just have to be
more subtle, and willing to receive an eye roll or two. If your eleven-
year-old is on the couch, less than inclined to join you and his
younger siblings in a board game, you can shift the mood by playfully
sitting on him. Again, this has to be done in a considerate way and fit
with his personality and mood, but a playfully apologetic "Oh, I'm
sorry. I didn't see you there" can at least draw a pretend-frustrated
"Daaaad" and, again, change the dynamics of the situation.

One reason this type of playfulness and humor can be effective
with kids—and adults as well, by the way—is that the brain loves
novelty. If you can introduce the brain to something it hasn't seen
before, something it didn't expect, it will give that something its attention. This makes sense from an evolutionary perspective: something that's different from what we usually see will pique our interest
on a primitive and automatic level. After all, the brain's first task is to
appraise any situation for safety. Its attention immediately goes to
whatever is unique, novel, unexpected, or different, so that it can as-
sess whether the new element in its environment is safe or not. The
appraisal centers of the brain ask, "Is this important? Do I need to
pay attention here? Is this good or bad? Do I move toward it or away

INSTEAD OF COMMANDING AND DEMANDING...

BE CREATIVE AND PLAYFUL

from it?" This attention to novelty is a key reason that humor and silliness can be so effective in a disciplinary moment. Also, a respectful sense of humor communicates the absence of threat, which allows our social engagement circuitry to engage, which in turn opens us up to connect with others. Creative responses to disciplinary situations prompt our kids' brains to ask these questions, become more receptive, and give us their full attention.

Creativity comes in handy in all kinds of other ways, too. Let's say your preschooler is using a word you don't like. Maybe she's saying things are "stupid." You've tried ignoring it, but you keep hearing the word. You've tried rephrasing it with a more acceptable synonym— "You're right, those swim goggles are just wacky, aren't they?"—but she keeps saying the goggles are stupid.

If ignoring and re-languaging don't turn out to be effective strategies, then instead of forbidding the word—you know how well that works—get creative. One gifted preschool director came up with an inspired way to address the use of the word. Anytime he heard a child say something was stupid, he would explain, in a matter-of-fact tone, that the word is really only meant to be used in a particular context: "'Stupid' is such a great word, isn't it? But I'm afraid you're using it wrong, my dear. You see, that's a very particular word that's really meant to be used only when talking to baby chickens. It's sort of a farm word. Let's come up with another term to use in this situation."

There are plenty of ways to approach a situation like this. You might suggest devising a code word that means "stupid," so that you two share a secret language that no one else understands. Maybe the new term could be "glooby" or some other fun word to say, or it could even be a hand signal you make up together. The point is that you find a way to creatively redirect your child toward behavior that will work better for everyone involved, and even give you a fun sense of connection.

Let's acknowledge one thing, though: sometimes you don't *feel* like being creative. It feels like it takes too much energy. Or maybe you're not too happy with your kids because of the way they're acting, so you're not exactly thrilled with the idea of mustering the energy to help them shift their mood or see things in a new light. In other words, sometimes you just don't want to be playful and fun. You want them to just get in the car seat without a song and dance! You want them to just put on their stinking shoes! You want them to just get their homework done, or turn off the video game, or stop fighting, or whatever!

We get it. Boy, do we get it.

However, compare the two options. The first is to be creative, which often demands more energy and goodwill than we can easily muster when we don't like the way our kids are acting. Ugh.

The other option, though, is to continue to have to participate in whatever battle the discipline situation has created. Double ugh. Doesn't it usually end up taking much more time and much more energy to engage in the battle? *The fact is, we can often completely avoid the battle by simply taking just a few seconds to come up with an idea that's fun and playful.*

So the next time you see trouble coming with your kids, or if there is a particular issue that you typically end up battling over, think about your two options. Ask yourself: "Do I really want the drama that's on the horizon?" If not, try playfulness. Be silly. Even if you don't feel like it, muster up the energy to be creative. Sidestep the drama that sucks the life out of you and takes the fun out of your relationship with your child. We promise, this option is more fun for everyone.

Redirection Strategy #8: Teach Mindsight Tools

The final redirection strategy we'll discuss is perhaps the most revolutionary. You'll recall that mindsight is all about seeing our own minds, as well as the minds of others, and promoting integration in our lives. Once kids begin to develop the personal insight that allows them to see and observe their own minds, they can then learn to use that insight to handle difficult situations.

We discussed this idea in detail in our previous book, *The Whole-Brain Child*, focusing on several Whole-Brain strategies parents can use to help their children integrate their brains and develop mindsight. As we've taught the fundamentals of that book to audiences of parents, therapists, and educators, we've further refined those ideas.

The overall point of this final redirection strategy is one that even small children can understand, although older kids can obviously grasp the message in more depth: *You don't have to get stuck in a negative experience. You don't have to be a victim to external events, or internal emotions. You can use your mind to take charge of how you feel, and how you act.*

We realize that this is an extraordinary promise to make. But we are enthusiastic about this approach because of how it has worked for so many people through the years. Parents really can teach their kids and themselves mindsight tools that will help them weather emotional storms and deal more effectively with difficult experiences, thus leading them to make better decisions and enjoy less chaos and drama when they are upset. *We can help our children increasingly have a say in how they feel, and in how they look at the world.* Not through some mysterious, mystical process available only to the gifted, but by using emerging knowledge about the brain and applying it in simple, logical, practical ways.

For example, you may have heard about the famous Stanford marshmallow experiment from the 1960s and 1970s. Young children were brought into a room one at a time, and a researcher invited them to sit down at a table. On the table was a marshmallow, and the researcher explained that he would leave the room for a few minutes. If the child resisted the temptation to eat the marshmallow while he was gone, he would give the child two marshmallows when he returned.

The results were predictably hilarious and adorable. Search online and you can view video of numerous replications of the study, which show children variously closing their eyes, covering their mouths, turning their back to the marshmallow, stroking it like a stuffed animal, slyly nibbling at the corners of the marshmallow, and so on.

Some children even grab the sugary treat and eat it before the researcher can finish delivering the instructions.

Much has been written about this study and follow-up experiments focusing on children's ability to delay gratification, demonstrate self-control, apply strategic reasoning, and so on. Researchers have found that kids who demonstrated the ability to wait longer before eating the marshmallow tended to have many improved life outcomes as they grew up, such as doing better in school, scoring higher on the SAT, and being more physically fit.

The application we want to highlight here is what a recent study revealed about how children could use mindsight tools to be more successful at delaying gratification. Researchers found that if they provided the kids with mental tools that gave them a perspective or strategy to assist in containing their impulse to eat the marshmallow—thus helping them manage their emotions and desires in that moment—the children were much more successful at demonstrating self-control. In fact, when the researchers taught the kids to imagine that it wasn't an actual marshmallow in front of them, but instead only a picture of a marshmallow, they were able to wait much longer than the kids who weren't given any strategies to help them wait! In other words, simply by using a simple mindsight tool, the children were able to more effectively manage their emotions, impulses, and actions.

You can do the same for your kids. If you've read *The Whole-Brain Child*, you know about the hand model of the brain. Here's how we introduced it in a "Whole-Brain Kids" cartoon for parents to read to their children.

WHOLE-BRAIN KIDS: Teach Your Kids About Their Downstairs and Upstairs Brain

YOUR DOWNSTAIRS BRAIN AND YOUR UPSTAIRS BRAIN

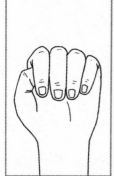

MAKE A FIST WITH YOUR HAND. THIS IS WHAT WE CALL A HAND MODEL OF YOUR BRAIN. RE- MEMBER HOW YOU HAVE A LEFT SIDE AND A RIGHT SIDE TO YOUR BRAIN? WELL, YOU ALSO HAVE AN UPSTAIRS AND A DOWN- STAIRS PART OF YOUR BRAIN.

THE UPSTAIRS BRAIN IS WHERE YOU MAKE GOOD DECISIONS AND DO THE RIGHT THING, EVEN WHEN YOU ARE FEELING REALLY UPSET.

NOW LIFT YOUR FINGERS A LITTLE BIT. SEE WHERE YOUR THUMB IS? THAT'S PART OF YOUR DOWN- STAIRS BRAIN, AND IT'S WHERE YOUR REALLY BIG FEELINGS COME FROM. IT LETS YOU CARE ABOUT OTHER PEOPLE AND FEEL LOVE. IT ALSO LETS YOU FEEL UPSET, LIKE WHEN YOU'RE MAD OR FRUS- TRATED.

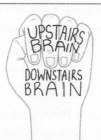

THERE'S NOTHING WRONG WITH FEELING UPSET. THAT'S NORMAL, ESPECIALLY WHEN YOUR UPSTAIRS BRAIN HELPS YOU CALM DOWN. FOR EXAMPLE, CLOSE YOUR FINGERS AGAIN. SEE HOW THE UPSTAIRS THINKING PART OF YOUR BRAIN IS TOUCHING YOUR THUMB, SO IT CAN HELP YOUR DOWNSTAIRS BRAIN EXPRESS YOUR FEELINGS CALMLY?

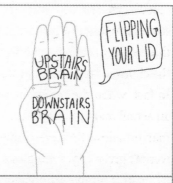

SOMETIMES WHEN WE GET REALLY UPSET, WE CAN FLIP OUR LID. RAISE YOUR FINGERS LIKE THIS. SEE HOW YOUR UPSTAIRS BRAIN IS NO LONGER TOUCHING YOUR DOWNSTAIRS BRAIN? THAT MEANS IT CAN'T HELP IT STAY CALM.

FOR EXAMPLE:

THIS IS WHAT HAPPENED TO JEFFREY WHEN HIS SISTER DESTROYED HIS LEGO TOWER. HE FLIPPED HIS LID AND WANTED TO SCREAM AT HER.

BUT JEFFREY'S PARENTS HAD TAUGHT HIM ABOUT FLIPPING HIS LID, AND HOW HIS UPSTAIRS BRAIN COULD HUG HIS DOWN-STAIRS BRAIN AND HELP HIM CALM DOWN. HE WAS STILL ANGRY, BUT INSTEAD OF SHOUTING AT HIS SISTER, HE WAS ABLE TO TELL HER HE WAS ANGRY AND ASK HIS PARENTS TO CARRY HER OUT OF HIS ROOM.

SO THE NEXT TIME YOU FEEL YOURSELF STARTING TO FLIP YOUR LID, MAKE A BRAIN MODEL WITH YOUR HAND. (REMEMBER IT'S A BRAIN MODEL, NOT AN ANGRY FIST!) PUT YOUR FINGERS STRAIGHT UP, THEN SLOWLY LOWER THEM SO THAT THEY'RE HUGGING YOUR THUMB. THIS WILL BE YOUR REMINDER TO USE YOUR UPSTAIRS BRAIN TO HELP YOU CALM THOSE BIG FEELINGS IN THE DOWN-STAIRS BRAIN.

Dan recently received an email from a school principal about a new kindergarten student who was struggling. The child's teacher had taught her class the hand model of the brain, and she saw immediate results:

Yesterday a teacher came to me very concerned about the behavior of a new kindergarten student. He had just come to our school, and he was crawling under tables and saying he hated everything. (He is living with a family member, as his mom is incarcerated, and now he's had to leave a teacher he really liked.)

Today our teacher retaught Brain-in-the-Hand. This was new to him. He was under the table most of the time while she taught. Soon after, he motioned to her, showed the flipped lid with his hand, and, on his own, went to the cool-off spot for a long time. (He almost fell asleep.)

When he finally got up, he approached her while she was teaching, pointed to his hand/brain with his lid closed, and joined the group.

After a bit she complimented him for his participation, and he said, "I know. I told you." And he pointed to his hand/brain with the lid closed.

It was a huge moment, and she and I celebrated for him that he must really have needed that language!

Later today I went in during choice time and played "restaurant" with him. At one point he took a single flower out of a vase and handed it to me. My heart melted. Yesterday his teacher was comparing him to a child who truly struggles. Today he's seeking every opportunity to connect with us. I'm so thankful that we're learning this.

What did this teacher do? She gave her student a mindsight tool. She helped him develop a strategy for understanding and expressing what was happening around and within him, so he could then make intentional choices about how to respond.

Another way to say it is that we want to help kids develop *a dual mode of processing the events that occur in their lives.* The first mode is all about teaching children to be aware of and simply sense their subjective experiences. In other words, when they're dealing with something difficult, we don't want them to deny that experience, or to squelch their emotions about it. We want them to talk about what's going on as they describe their inner experience, communicating what they're feeling and seeing in that moment. That's the first mode of processing: to simply acknowledge and be present with the experience. This teacher, in other words, didn't want this little boy to deny how he was feeling. His feeling was his experience, and this "experiencing mode" is all about simply sensing inner subjective experience as it is happening.

But also we want our kids to be able to *observe* what's going on within them, and how the experience is impacting them. Brain studies reveal that we actually have two different circuits—an experiencing circuit and an observing circuit. They are different, but each is important, and integrating them means building both and then linking them. We want our kids to not only feel their feelings and sense their sensations, but also to be able to *notice* how their body feels, to be able to *witness* their own emotions. We want them to pay attention to their emotions ("I'm noticing that I'm feeling kind of sad," or "My frustration isn't grape-size right now; it's like a watermelon!"). We want to teach them to survey themselves, and then problem-solve based on this awareness of their internal state.

That's what this boy did. He both lived in his experience *and* observed it. This allowed him to *own* what was going on. He had the perspective to be able to *observe* his experience as he was experiencing it. He could bear *witness* to the unfolding of experience, not just be in the experience. And then he could *narrate* what had happened, using language to express to others and to himself an understanding of what was going on. Using the hand model as his tool, he surveyed himself and recognized that he had "flipped his lid," and he took steps in response, thus changing his internal state. Then when he was back in control of his emotions, he rejoined the group.

We see kids and parents in our work who become stuck in an experience they're dealing with. Of course they need to deal with what's happened to them. But that's only one mode of processing. They also need to look at and think about what's going on. They need to use mindsight tools to become aware of and observe, almost like a reporter, what is happening. One way to explain it is that we want them to be the actor, experiencing the scene in the moment, but also to be the director, who watches more objectively and can, from outside the scene, be more insightful about what's taking place on camera.

When we teach kids to be both actor and director—to embrace the experience and also to survey and observe what's happening within themselves—we give them important tools that help them take charge of how they respond to situations they're faced with. It allows them to say, "I hate tests! My heart is pounding, and I'm starting to freak out!" but then also to observe, "That's not weird. I really want to do well on it. But I don't have to freak out. I just need to skip that TV show tonight and put in some extra study time."

Again, this is about teaching kids that they don't have to be stuck in an experience. They can also be observers and therefore change agents. Let's say, for example, that the child described above remains overly concerned about tomorrow's test. He begins a cascade of worrying that takes him into a spiral of panic about the test and his semester grade, and what that might mean in terms of graduating with the right GPA to get into a good college.

This would be a great time for his parents to teach him that he can change his emotions and his thinking by moving his body, or simply by altering his physical posture. In *The Whole-Brain Child*, we call this particular mindsight tool the "move it or lose it" technique. The boy's parents could have him sit "like a noodle," completely relaxed and "floppy," for a couple of minutes. They all could then observe together how his feelings, thoughts, and body began to feel different. (It really is amazing how effective this particular strategy can be when we're tense.) Then they could go back and talk about the exam from an "unstuck place" where he could see that he had some options.

There are limitless ways you can teach your kids about the power of the mind. Explain the concept of shark music, and have a conversation about what experiences from the past might be impacting their decision making. Or explain the river of well-being. Show them the picture from Chapter 3, and walk them through a discussion of a recent experience when they were especially chaotic or rigid. Or when they are feeling scared about something, tell them, "Show me what your body looks like when you're brave, and let's see what that feels like." Recent studies are suggesting that simply holding our bodies in various postures can actually shift our emotions, along with the way we view the world.

Opportunities to teach mindsight tools are everywhere. In the car, when your nine-year-old is upset about an important shot she missed in her basketball game, direct her attention to the splotches on your windshield. Say something like, "Each spot on the windshield is something that has happened or will happen this month. This one here is your basketball game. That's real, and I know you're upset. I'm glad you're able to be aware of your feelings. But look at all the other splotches on the windshield. This one here is the party this weekend. You're pretty excited about that, aren't you? And that one next to it represents your math grade from yesterday. Remember how proud you felt?" Then continue the conversation, putting the missed shot into context with her other experiences.

The point of an exercise like this isn't to tell your daughter not to worry about her basketball game. Not at all. We want to *encourage* our kids to feel their feelings, and to share them with us. The sensing mode that lets us experience directly is an important mode of processing. But along the way, we want to give them perspective and help them understand that they can focus their attention on other aspects of their reality. This comes from having our *observing* circuits well developed, too, not just our *sensing* circuits. It's not a matter of one or the other. Both are important, and together they make a great team. That's one way we can help our kids develop integration by differentiating and then linking their sensing and absorbing capacities.

Having built both circuits, our kids can use their minds to think about things other than what's upsetting them in a particular moment, and as a result, they see the world differently and feel better. When we teach our kids mindsight tools, we give them the gift of being able to regulate their emotions, rather than being ruled by them, so they don't have to remain victims of their environment or their emotions.

The next time a discipline opportunity comes up in your house, introduce your kids to some mindsight tools. Or use one of the other redirection strategies we've presented here. You might have to try several different approaches. No one strategy will apply in every situation. But if you work from a No-Drama, Whole-Brain perspective that first connects, then redirects, you can more effectively achieve the primary goals of discipline: gaining cooperation in the moment and building your children's brains so they can be kind and responsible people who enjoy successful relationships and meaningful lives.

On Magic Wands, Being Human, Reconnection, and Change: Four Messages of Hope

We've emphasized throughout this book that No-Drama Discipline allows for a much calmer and more loving disciplinary interaction. We've also said that a No-Drama, Whole-Brain approach not only is better for your children, their future, and your relationship with them, but actually makes discipline more effective and your life easier as well, since it increases the cooperation you'll receive from your kids.

Still, even with the best ambitions and the most intentional methods, sometimes everyone walks away from a disciplinary interaction feeling angry, confused, and frustrated. In our closing pages, we want to offer four messages of hope and solace for those difficult moments we all inevitably face at one time or another as we discipline our children.

First Message of Hope: There Is No Magic Wand

One day Tina's seven-year-old became furious with her because she told him it wouldn't work that day to invite a friend over to play. He stormed off to his room and slammed the door. Less than a minute later, she heard the door open, then slam again.

Here's how Tina tells the story.

I went to check on my son, and taped to the outside of his
door, I saw this picture.

(You can see from the drawing below that he regularly uses his
artistic talents to communicate his feelings about his parents.)

I went into his room and saw what I knew I'd see: a child-size lump under the covers on his bed. I sat next to the lump and put my hand on what I assumed was a shoulder, and suddenly the lump moved away from me, toward the wall. From beneath the covers, my son cried out, "Get away from me!"

At times like this I can become childish and drop down to my child's level. I've even been known to say things like, "Fine! If you won't let me cut that toenail that's hurting, you can stay in pain all week!"

But this particular day, I maintained control and handled myself pretty well, trying to address the situation from a Whole-Brain perspective. I first tried to connect by acknowledging his feelings: "I know it makes you mad that Ryan can't come over today."

His response? "Yes, and I hate you!"

I stayed calm and said, "Sweetie, I know this is upsetting, but there's just not time to have Ryan over today. We're meeting your grandparents for dinner in a little while."

In response, he curled tighter and moved as far away from me as possible. "I said get away from me!"

I went through a series of strategies, the ones we've been discussing in the previous chapters. I comforted, using non-verbal connection. I tried to relate to his changing, changeable, complex brain. I chased the why and thought about the how of my communication. I validated his feelings. I tried to engage in a collaborative dialogue and reframed my no, offering a playdate the next day. But at that moment, he couldn't calm down and wasn't ready to let me help him in any way. No amount of connection did the trick.

Moments like these highlight a reality that's important for parents to understand: sometimes there's just nothing we can do to "fix" things when our kids are having a hard time. We can work to stay calm and loving. We can be fully present. We can access the full measure of our creativity. And still, we may not be able to make things better right away. Sometimes all we have to offer is our presence as our children move through the emotions. When kids clearly communicate that they want to be alone, we can respect what they feel they need in order to calm down.

This doesn't mean we'd leave a child crying alone in his room for long periods of time. And it doesn't mean we don't keep trying different strategies when our child needs our help. In Tina's case, she ended up sending her husband into her son's room, and the change of dynamic helped him begin to calm some, so that later he and his mom could come back together and talk about what happened. But for a few minutes, all Tina could do was say, "I'm here if you need me," then leave him in his room for a few minutes, shut the door with the anti-Mom sign on it, and let him ride it out the way he needed to, on his own timing and in his own way.

The same goes for sibling conflict. The ideal is to help each sibling return to a good state of mind, then work with them, individually or together, and teach them good relational and conversation skills. But there are times this just isn't possible. If even just one of them is emotionally dysregulated, it can prevent anything like a peaceful resolution, since reactivity is trumping receptivity. Sometimes the best you can do is separate them until you can all come together again once everyone has calmed down. And if cruel fate decrees that you're all trapped in the minivan when the conflict erupts, you may just need to explicitly acknowledge that things are not going well and turn up the music. In doing so, you're not surrendering. You're just acknowledging that at this moment, effective discipline isn't going to happen. In cases like this, you can say, "This isn't a good time for us to talk this through. You're both mad, and I'm mad, so let's just listen to some Fleetwood Mac." (OK, maybe that's not the best choice in music to win your kids over, but you get the idea.)

We, Dan and Tina, are both trained child and adolescent psychotherapists who write books about parenting. People come to us for advice on how to handle problems when their kids are struggling.

And we want to make it clear that for us, like you, there are times when there just isn't a magic wand we can wave to magically transport our kids to peace and happiness. Sometimes the best we can do is to communicate our love, be available when they do want us close, and then talk about the situation when they're ready. It's just like the Serenity Prayer says: "May I have the courage to change the things I can, the serenity to accept the things I can't, and the wisdom to know the difference."

So that's our first message as we conclude the book: sometimes there's no magic wand. And it doesn't make you a bad parent if you do your best, and your child stays upset.

Second Message of Hope: Your Kids Benefit Even When You Mess Up

Just as it doesn't make you a bad parent if your discipline techniques aren't always effective in the moment, you're also not a bad parent if you make mistakes on a regular basis. What you are is human.

The fact is that none of us are perfect, especially when it comes time to deal with our kids' behavior. Sometimes we handle ourselves well and feel proud of how loving, understanding, and patient we remain. At other times, we lower ourselves to our kids' level and resort to the childishness that upset us in the first place.

Our second message of hope is that when you respond to your kids from a less-than-optimal place, you can take heart: most likely you're still providing them with all kinds of valuable experiences.

For example, have you ever found yourself so frustrated with your kids that you call out, a good bit louder than you need to, "That's it! The next one who complains about where they're sitting in the car can walk!" Or maybe, when your eight-year-old pouts and complains all the way to school because you made her practice the piano, you deliver these sarcastic and biting words as she departs the minivan: "I hope you have a great day, now that you've ruined the whole morning."

Obviously, these aren't examples of optimal parenting. And if you're like us, you can be hard on yourself for the times you don't handle things like you wish you had.

So here's hope: those not-so-great parenting moments are not necessarily such bad things for our kids to have to go through. In fact, they're actually incredibly valuable.

Why? Because our messy, human, parental responses give kids opportunities to deal with difficult situations and therefore develop new skills. They have to learn to control themselves even though their parent isn't doing such a great job of controlling herself. Then they get to see you model how to apologize and make things right. They experience that when there is conflict and argument, there can be repair, and things become good again. This helps them feel safe and not so afraid in future relationships; they learn to trust, and even expect, that calm and connection will follow conflict. Plus, they learn that their actions affect other people's emotions and behavior. Finally, they see that you're not perfect, so they won't expect themselves to be, either. That's a lot of important lessons to learn from one parent's loud, impulsive declaration that he's sending back all the presents because his kids complained about having to help put up the holiday decorations.

Abuse, of course, is different, whether physical or psychological. Or if you're significantly harming the relationship or scaring your child, then the experience can result in substantially harmful effects. These are toxic ruptures, and ruptures without repair. If you find yourself in that situation repeatedly, you should seek the help of a professional right away in order to make whatever changes are necessary so that your children are safe and know that they are protected.

But as long as you nurture the relationship and repair with your child afterward (more about that below), then you can cut yourself some slack and know that even though you might wish you'd done things differently, you've still given your child a valuable experience, by learning the importance of repair and reconnection.

We hope it's obvious that we're not saying that parents should intentionally rupture a connection or that we shouldn't aim for the best when we respond to our kids in a high-stress situation (or any other time). The more loving and nurturing we can be, the better. Those non-ideal moments of non-optimal interactions will happen to all of us, even those of us who write books on this subject. We're just saying that we can offer grace and forgiveness to ourselves when we're not acting as we'd like to, because even those situations provide moments of value as well. Having a goal, an intention in mind, is important. And being kind to ourselves, having self-compassion, is essential not only to create an internal sanctuary, but also to offer our children a role model for being kind to themselves as well as to others. These experiences with us give our kids opportunities to learn important lessons that will prepare them for future conflict and relationships, and even teach them how to love. How's that for hope?

Third Message of Hope: You Can Always Reconnect

There's no way we can avoid experiencing conflict with our kids. It's going to happen, sometimes multiple times per day. Misunderstandings, arguments, conflicting desires, and other breakdowns in communication will lead to a rupture in the relationship. Ruptures can result from conflict around a limit that you're setting. Maybe you decide to enforce a bedtime or keep your child from seeing a movie you've decided isn't good for him. Or maybe your daughter thinks you're taking her sister's side in an argument, or she gets frustrated that you won't play another game of Chutes and Ladders.

Whatever the reason, ruptures occur. Sometimes they are bigger, sometimes smaller. But there's no way to avoid them. Each child presents a unique challenge to maintaining attuned connection, one that depends on our own issues, on our child's temperament, on the match between our history and our child's characteristics, and on whom our child may remind us of in our own un-worked-through past.

In most of our adult relationships, if we mess up, we eventually own up to it, or address it in some way, and then make amends. But many parents, when it comes to their relationship with their child, just ignore the rupture and never address it. This can be confusing and hurtful for children, just like it can be for adults. Can you imagine someone you care about being reactive and talking to you really rudely, then never bringing it up again and just pretending it never happened? That wouldn't feel great, would it? It's the same for our kids.

What's key, then, is that you repair any breach in the relationship as quickly as possible. You want to restore a collaborative, nurturing connection with your child. Ruptures without repair leave both parent and child feeling disconnected. And if that disconnection is prolonged—and especially if it's associated with your anger, hostility, or rage—then toxic shame and humiliation can grow in the child, damaging her emerging sense of self and her state of mind about how relationships work. It's therefore vital that we make a timely reconnection with our kids after there's been a rupture.

It's our responsibility as parents to do this. Maybe we reconnect by granting forgiveness, or by asking for it ("I'm sorry. I think I was just reacting because I'm extra tired today. But I know I didn't handle myself very well. I'll listen if you want to talk about what that was like for you"). Maybe laughter's involved, maybe tears ("Well, that didn't go very well, did it? Anyone care to play back for me how crazy I was?"). Maybe there's just a quick acknowledgment ("I didn't handle that how I would have liked. Will you forgive me?"). However it happens, make it happen. By repairing and reconnecting as soon as we can, and in a sincere and loving manner, we reconnect and send the message that the relationship matters more than whatever caused the conflict. Plus, in reconnecting with our kids, we model for them a crucial skill that will allow them to enjoy much more meaningful relationships as they grow up.

So that's the third message of hope: we can always reconnect. Even though there's no magic wand, our kids will eventually soften and calm down. They'll eventually be ready to sense our positive

REPAIR A RUPTURE ASAP

intentions and receive our love and comfort. When they do, we reconnect. And even though we're going to mess up as parents over and over again because we are human, we can always go to our kids and repair the breach.

In the end, then, it all comes back to connection. Yes, we want to redirect. We want to teach. Our children need us to help them learn how to focus their desires in positive ways; how to recognize and deal with limits and boundaries; how to discover what it means to be human and to be moral, ethical, empathic, kind, and giving. So yes, redirection is crucial. But ultimately, it's your relationship with your child that must always stay at the forefront of your mind. *Put any particular behavior on the back burner, and keep your relationship with your child always on the front burner.* Once that relationship has been ruptured in any way, reconnect as soon as possible.

Fourth Message of Hope:
It's Never Too Late to Make a Positive Change

Our final message for you is the most hopeful of all: it's never too late to make a positive change. Having read this book, you may now feel that your discipline approach up to this point has at least partially run counter to what's best for your children. Perhaps you feel that you're undermining your relationship with them by the way you discipline. Or maybe you realize that you're overlooking and missing out on opportunities to build the parts of their brains that will help them achieve optimal growth. You might now see that you're using disciplinary strategies that are simply not effective, are just contributing to more drama and frustration in your family, and are actually keeping you from enjoying your kids because you end up having to deal with the same behaviors over and over.

If any of that's the case, have hope. It's not too late. Neuroplasticity, as we've said, shows us that the brain is amazingly changeable and adaptive across a lifetime. You can change the way you discipline at any age—yours or your child's. No-Drama Discipline shows you how. Not by offering a formula to follow. Not by providing a magic wand that will solve every problem and make you a parent who never misses the mark. The hope comes in that you now have principles that can guide you toward disciplining your children in ways you can feel good about. You now have access to strategies that actually sculpt the brain in positive ways, allow your kids to be emotionally intelligent and make good choices, strengthen your relationship with them, and help them become the kind of people you want them to be.

When you respond to your kids with connection—even and especially when they do something that frustrates you—you put your primary focus not on punishment or obedience, but on honoring both your child and the relationship. So the next time your toddler throws a tantrum, your second grader punches his sister, or your middle schooler talks back, you can choose to respond in a No-Drama, Whole-Brain fashion. You can begin with connection, then move on to redirection strategies that teach kids personal insight, relational empathy, and the importance of taking responsibility for the times they mess up.

Along the way, you can be more intentional about how you activate certain circuits of your kids' brains. Neurons that fire together wire together. The circuitry that is repeatedly activated will be strengthened and further developed. So the question is, which part of your kids' brains do you want to strengthen? Discipline with harshness, shouting, arguments, punishment, and rigidity, and you'll activate the downstairs, reactive part of your child's brain, strengthening that circuitry and priming it to be easily activated. Or discipline with calm, loving connection, and you'll activate the reflective, receptive, regulating mindsight circuitry, strengthening and developing the upstairs section of the brain to create insight, empathy, integration, and repair. *Right now, in this moment, you can commit to giving your children these valuable tools. You can help them develop this increased capacity to regulate themselves, to make good choices, and to handle themselves well—even in challenging times, and even when you're not around.*

You're not going to be perfect, and you're not going to discipline from a No-Drama, Whole-Brain perspective every time you get the chance. Neither do we. Nobody does.

But you can decide that you'll take steps in that direction. And every step you take, you'll give your kids the gift of a parent who is increasingly committed to their lifelong success and happiness, and to making them happy, healthy, and fully themselves.

Further Resources

CONNECT AND REDIRECT REFRIGERATOR SHEET

No-Drama Discipline
by Daniel J. Siegel, M.D., and Tina Payne Bryson, Ph.D.

FIRST, CONNECT

- **Why connect first?**
  - *Short-term benefit:* It moves a child from reactivity to receptivity.
  - *Long-term benefit:* It builds a child's brain.
  - *Relational benefit:* It deepens your relationship with your child.

- **No-Drama connection principles**
  - *Turn down the "shark music":* Let go of the background noise caused by past experiences and future fears.
  - *Chase the why:* Instead of focusing only on behavior, look for what's *behind* the actions: "Why is my child acting this way? What is my child communicating?"
  - *Think about the how:* What you say is important. But just as important, if not more important, is *how* you say it.

- **The No-Drama connection cycle: help your child feel felt**
  - *Communicate comfort:* By getting below your child's eye level, then giving a loving touch, a nod of the head, or an empathic look, you can often quickly defuse a heated situation.
  - *Validate:* Even when you don't like the behavior, acknowledge and even embrace feelings.
  - *Stop talking and listen:* When your child's emotions are exploding, don't explain, lecture, or try to talk her out of her feelings. Just listen, looking for the meaning and emotions your child is communicating.
  - *Reflect what you hear:* Once you've listened, reflect back what you've heard, letting your kids know you've heard them. That leads back to communicating comfort, and the cycle repeats.

THEN, REDIRECT

- **1-2-3 discipline, the No-Drama way**
  - One definition: Discipline is teaching. Ask the three questions:
    1. Why did my child act this way? (What was happening internally/emotionally?)
    2. What lesson do I want to teach?
    3. How can I best teach it?
  - Two principles:
    1. Wait until your child is ready (and you are, too).
    2. Be consistent but not rigid.
  - Three mindsight outcomes:
    1. *Insight:* Help kids understand their own feelings and their responses to difficult situations.
    2. *Empathy:* Give kids practice reflecting on how their actions impact others.
    3. *Repair:* Ask kids what they can do to make things right.

- **No-Drama redirection strategies**
  - Reduce words
  - Embrace emotions
  - Describe, don't preach
  - Involve your child in the discipline
  - Reframe a no into a yes with conditions
  - Emphasize the positive
  - Creatively approach the situation
  - Teach mindsight tools

WHEN A PARENTING EXPERT LOSES IT

You're Not the Only One

J ust because we write books about parenting and discipline
doesn't mean there aren't times when we mess up with our own
kids. Here are two stories—one from each of us—that, while pretty
funny in retrospect, show that the reactive brain can get us all.

Dan's "Crepes of Wrath" Moment (adapted from Dan's book *Mindsight*)

One day my thirteen-year-old son, my nine-year-old daughter, and I
stopped into a small shop for a snack after a movie. My daughter said
she wasn't hungry, and so my son ordered a small crepe for himself
from the counter and we sat down. The simple crepe arrived, aromas
wafting from the open kitchen behind the counter where my son had
placed his order. After my son took his first forkful of crepe, my
daughter asked if she could try some. My son looked at the small
crepe and said that he was hungry and she could order her own. It
was a reasonable suggestion, I thought, so I offered to get another
crepe for her—but she said she wanted only a small bite to see how it
tasted. That also seemed reasonable, so I suggested that my son share
a piece with his sister.

If you have more than one child at home, or if you've grown up with a brother or sister, you may be very familiar with the game of sibling chess, an ever-present strategy match composed of movements aimed to assert power and achieve parental recognition and approval. But even if this was not such a sibling assertion game, the small cost of buying the additional crepe from this little family-run crepe shop would have been quite a simple one to pay to avoid what you may guess was about to happen. Instead of making the purchase, I made a parental blunder and took sides in this sibling game. I firmly insisted that my son share his crepe with his sister. If this was not a sibling chess match before, it certainly became one after I stepped into their interaction.

"Why don't you just give her a small piece so she can see how it tastes?" I urged.

He looked at me, then at his crepe, and with a sigh he gave in. Even as a young teenager he was still listening to me. Then, using his knife like a scalpel, he extracted the smallest piece of crepe you can imagine, one you'd almost need tweezers to pick up. Under other circumstances, I might have laughed and seen this as a creative move in the sibling chess game.

My daughter took the specimen, placed it on her napkin, and said that it was too small. And that it was "the burnt part." Another great younger-sister move.

An outsider looking in at us at the table may have seen nothing out of the ordinary: a dad and his two animated kids out for some food. But inside, I was about to explode. When the bantering continued, turning into a full-blown argument, something inside me shifted. My head began to spin, but I told myself that I'd remain calm and appeal to reason. I could feel my face tense up, my fists get taut, and my heart begin to beat faster, but I tried to ignore these signals that my downstairs brain was hijacking the upstairs. That was it for me.

Feeling overwhelmed by the ridiculousness of the whole encounter, I got up, took my daughter's hand, and went outside to wait on the sidewalk in front of the shop until my son finished his crepe. A few minutes later he emerged and asked why we had left. As I stormed off toward the car, my daughter in tow and my son hurrying to keep up, I told them that they should learn to share their food with each other. He pointed out in a matter-of-fact tone that he did give her a piece, but by then I was boiling over with frustration, and at that point there was no turning off the heat under the kettle. We got to the car, and, fired up, I ignited the engine and away we went toward home. They had been normal siblings out for movies and a snack. I became a father out of my mind.

I couldn't let it go. Sitting next to me in the passenger seat, my son countered everything I came up with by some rational, measured response, as any teenager would do. In fact, he seemed quite adept at staying calm as he dealt with his now irrational father.

In that state, I became more and more irate, eventually resorting to cursing, calling him names, and even threatening to take away his beloved guitar—all inappropriate consequences for things he didn't even do.

I'm not proud to tell you any of this. But Tina and I do feel that since such explosive episodes are quite common, it is essential that we acknowledge their existence and help each other understand how mindsight can diminish their negative impact on our relationships and on our world. In our shame, we often try to ignore that a meltdown has occurred. But if we own the truth of what has happened, we can not only begin to repair the damage—which can be quite toxic to ourselves as well as to others—but actually decrease the intensity of such events and the frequency with which they occur.

So when I got home, I realized that I needed to calm down and connect with my son. I knew repair was crucial, but my vital signs were through the roof, and I had to bring them into balance before doing anything else. Knowing that being outside and exercising could help alter my state of mind, I went skating with my daughter, during which time she helped me regain mindsight. I achieved more personal insight (recognizing that I reacted to my son the way I did at least partially because I was unconsciously identifying him with my own older brother) and empathy for how my son experienced our encounter.

When I finally cooled down after talking and skating and reflecting, I went to my son's room and asked if we could talk. I said that I thought I had gone off the deep end, and that it would be helpful for us to discuss what had happened. He told me that he thought I was too protective of his sister. He was absolutely right. Although the embarrassment of having become irrational created an urge to speak up to defend myself and my reactions, I just kept quiet. My son went on to tell me that my getting "upset" was unnecessary because he really hadn't done anything wrong. He was right. Again I felt a defensive urge to lecture him about sharing. But I reminded myself to remain reflective and focus on my son's experience, not mine. The essential stance here was not to judge who was right, but to be accepting and receptive to him. You can imagine that this all required mindsight, for sure. I was thankful my prefrontal region was back at work.

After listening to him, I acknowledged that I had in fact taken his sister's side (unfairly), that I could see how this felt unjust to him, and that my explosion seemed irrational—because in fact it was. As an explanation—not an excuse—I let him know what had happened in my mind, seeing him as a symbol of my brother, so that we both could make sense of the whole encounter. Even though I probably looked awkward and clumsy in his teenage mind, I could tell that he knew my commitment to our relationship was deep and my effort to repair the damage was genuine. My mindsight had returned, our two minds connected again, and our relationship was back on track.

Tina Threatens an Amputation

When my oldest child was three years old, he hit me one day. As a young and idealistic parent who, at that time, believed that my best alternative was to have a rational conversation with a three-year-old in which he would magically see things from my point of view, I guided him to the bottom of our stairway, sat next to him, and smiled.

I lovingly (and naively) said, "Hands are for helping and loving, not for hurting."

While I was uttering this truism, he hit me again.

So I tried the empathy approach. Still naive, my voice perhaps sounding a bit less loving, I said, "Ouch! That hurts Mommy. Be gentle with my body."

At which point he hit me again.

I then tried a more firm approach: "Hitting is not OK. We don't hit. If you're mad, you need to use your words."

Yup, you guessed it. He hit me again.

I was lost. I felt I needed to up the ante, but I didn't know how. In my most powerful voice, I said, "Now you're in time-out at the *top* of the stairs." (The technical, scientific term for this parenting strategy is "Flying by the seat of your pants." Not exactly intentional parenting.)

I marched him to the top of our stairs. He was probably thinking, "Cool! We've never done this before. . . . I wonder what will happen next if I keep hitting her?"

At the top of the stairs, I bent over at the waist, my finger wagging, and said, *"No more hitting!"*

He didn't hit me again.

He kicked me in the shin.

(As he points out these days when we retell the story, he was technically obeying my no-hitting instructions.)

At this moment virtually all of my self-control was gone, as were any viable options I could think of. I grabbed his arm and pulled him into my room at the top of the stairs, yelling, "Now you're in time-out in Mommy and Daddy's room!"

Again, I had no strategy, no plan or approach. And as a result, my young son continued to escalate the situation while his increasingly red-faced mother yanked him from location to location in the house.

By this point I was by turns cajoling, scolding, commanding, reacting, and reasoning (waaaay too much talking): "You may not hurt Mommy. Hitting and kicking are not how we do things in our family. . . . Blah blah blah . . ."

And that's when he made his biggest mistake. He stuck out his tongue at me.

In response, my rational, empathic, responsible, problem-solving upstairs brain was hijacked by my primitive, reactive, downstairs brain, and I yelled, *"If you stick that tongue out one more time, I'm going to rip it out of your mouth!"*

In case you're wondering, neither Dan nor I recommend in any circumstance threatening to remove any of your children's body parts. This was not a good parenting moment.

And it wasn't effective discipline, either. My son dropped to the ground, crying. I'd scared him, and he kept saying, "You're a mean mommy!" He wasn't thinking about his own behavior at all—he was solely focused on *my* misbehavior.

What I did next was probably the only thing I did right in the whole interaction, and it's essential each time we have these types of ruptures in our relationship with our children: I repaired with him. I immediately realized how awful I'd been in that reactive, angry moment. If anyone else had treated my child as I just had, I would've come unglued. I knelt down and joined my young son on the floor, held him close, and told him how sorry I was. I let him talk about how much he didn't like what had just occurred. We retold the story to make sense of it for him and I comforted him.

I usually get big laughs when I tell this story because parents so identify with this type of a moment, and I think they enjoy hearing that a parenting expert can totally lose it, too. As I explain to my audiences, we need to be patient, understanding, and forgiving—not only with our children, but with ourselves as well. (People always ask what I would do differently now. See Chapter 6, where we discuss addressing toddler misbehavior in four steps—with illustrations!)

Though these stories are a bit embarrassing to relate, we offer them as (yes, humorous) evidence that we are all potentially prone to such downstairs dis-integrations when we lose control and handle ourselves poorly. Episodes like these shouldn't become a regular occurrence, though. If you find yourself repeatedly losing it in intense ways, we recommend that you consider seeking professional help to assist you in making sense of your own emotional needs or woundings that may be contributing to frequently reactive ways of relating to your children. But if you go down the low road only every so often, as most of us do, that's just part of parenting. The key is recognizing when these moments happen, putting an end to them as quickly as possible to minimize the hurt they cause, and then making a repair. We need to regain what was truly lost—mindsight—and then use insight and empathy to reconnect with ourselves and repair with those for whom we care so deeply.

Our Discipline Approach in a Nutshell

You are an important person in the life of our child or children. You're helping determine who they're becoming by shaping their hearts, their character, and even the structures of their brains! Because we share this incredible privilege and responsibility of teaching them how to make good choices and how to be kind, successful human beings, we want to also share with you how we handle behavioral challenges, in hopes that we can work together to give our children a consistent, effective experience when it comes to discipline.

Here are the eight basic principles that guide us:

1. *Discipline is essential.* We believe that loving our kids and giving them what they need includes setting clear and consistent boundaries and holding high expectations for them—all of which helps them achieve success in relationships and other areas of their lives.

2. *Effective discipline depends on a loving, respectful relationship between adult and child.* Discipline should never include threats or humiliation, cause physical pain, scare children, or make them

feel that the adult is the enemy. Discipline should feel safe and loving to everyone involved.

3. *The goal of discipline is to teach.* We use discipline moments to build skills so kids can handle themselves better now and make better decisions in the future. There are usually better ways to teach than giving immediate consequences. Instead of punishment, we encourage cooperation from our kids by helping them think about their actions, and by being creative and playful. We set limits by having a conversation to help develop awareness and skills that lead to better behavior both today and tomorrow.

4. *The first step in discipline is to pay attention to kids' emotions.* When children misbehave, it's usually the result of not handling big feelings well and not yet having the skills to make good choices. So being attentive to the emotional experience *behind a behavior* is just as important as the behavior itself. In fact, science shows that addressing kids' emotional needs is actually the most effective approach to changing behavior over time, as well as developing their brains in ways that allow them to handle themselves better as they grow up.

5. *When children are upset or throwing a fit, that's when they need us most.* We need to show them we are there for them, and that we'll be there for them at their absolute worst. This is how we build trust and a feeling of overall safety.

6. *Sometimes we need to wait until children are ready to learn.* If kids are upset or out of control, that's the worst time to try to teach them. Those big emotions are evidence that our children need us. Our first job is to help them calm down, so they can regain control and handle themselves well.

7. *The way we help them be ready to learn is to connect with them.* Before we redirect their behavior, we connect and comfort. Just like we soothe them when they are physically hurt, we do the same when they're emotionally upset. We do this by validating their feelings and by giving them lots of nurturing empathy. Before we teach, we connect.

8. *After connecting, we redirect.* Once they've felt that connection with us, kids will be more ready to learn, so we can effectively redirect them and talk with them about their behavior. What do we hope to accomplish when we redirect and set limits? We want our kids to gain insight into themselves, empathy for others, and the ability to make things right when they make mistakes.

For us, discipline comes down to one simple phrase: *Connect and redirect.* Our first response should always be to offer soothing connection; then we can redirect behaviors. *Even when we say no to children's behavior, we always want to say yes to their emotions, and to the way they experience things.*

TWENTY DISCIPLINE MISTAKES

Even Great Parents Make

Because we're *always* parenting our children, it takes real effort to look at our discipline strategies objectively. Good intentions can be replaced by less-than-effective habits quickly, and that can leave us operating blindly, disciplining in ways that might not bring out our best—or the best in our children. Here are some common discipline mistakes made by even the best-intentioned, most well-informed parents. These mistakes crop up when we lose sight of our No-Drama, Whole-Brain goals. Keeping them in mind can help us to avoid them or to step back when we start heading down the low road.

1. Our discipline becomes consequence-based instead of teaching-based.

The goal of discipline is not to make sure that each infraction is immediately met with a consequence. The real goal is to teach our children how to live well in the world. But many times we discipline on autopilot, and we focus so much on the consequences that those become the end goal, the entire focus. So when you discipline, ask yourself what your real objective is. Then find a creative way to teach that lesson. You can probably find a better way to teach it without even using consequences at all.

2. We think that if we're disciplining, we can't be warm and nurturing.

It really is possible to be calm, loving, and nurturing while disciplining your child. In fact, it's important to combine clear and consistent boundaries with loving empathy. Don't underestimate how powerful a kind tone of voice can be as you have a conversation with your child about the behavior you want to change. Ultimately, you're trying to remain strong and consistent in your discipline while still interacting with your child in a way that communicates warmth, love, respect, and compassion. These two aspects of parenting can and should co-exist.

3. We confuse consistency with rigidity.

Consistency means working from a reliable and coherent philosophy so that our kids know what we expect of them. It doesn't mean maintaining an unswerving devotion to some sort of arbitrary set of rules. So at times you might make exceptions to the rules, turn a blind eye to some sort of minor infraction, or cut your child some slack.

4. We talk too much.

When kids are reactive and having a hard time listening, we often need to just be quiet. When we talk and talk at our upset children, it's usually counterproductive. We're just giving them a lot of sensory input that can further dysregulate them. Instead, use more nonverbal communication. Hold them. Rub their shoulders. Smile or offer empathic facial expressions. Nod. Then, when they begin to calm down and are ready to listen, you can redirect by bringing in the words and addressing the issue on a more verbal, logical level.

5. We focus too much on the behavior and not enough on the *why* behind the behavior.

Any good doctor knows that a symptom is only a sign that something else needs to be addressed. Children's misbehavior is usually a symptom of something else. It will keep occurring if we don't connect with our kids' feelings and their subjective experiences that lead to the behavior. The next time your child acts out, put on your Sherlock Holmes hat and look *through* the behavior to see what feelings—curiosity, anger, frustration, exhaustion, hunger, and so on—might be causing the behavior.

6. We forget to focus on *how* we say what we say.

What we say to our kids matters. Of course it does. But just as important is *how* we say it. Although it's not easy, we want to aim for being kind and respectful every time we communicate with our kids. We won't always be able to hit this mark, but that should be our goal.

7. We communicate that our kids shouldn't experience big or negative feelings.

When your child reacts intensely when something doesn't go his way, do you ever shut down that reaction? We don't mean to, but parents can often send the message that we're interested in being with our kids only if they're happy, and not when they're expressing negative emotions. We may say things like, "When you're ready to be nice, then you can rejoin the family." Instead, we want to communicate that we will be there for them, even at their absolute worst. Even as we say no to certain behaviors or to how certain feelings get expressed, we want to say yes to our kids' emotions.

8. We overreact, so our kids focus on our overreaction, not their own actions.

When we overshoot the mark with our discipline—if we're punitive, or we're too harsh, or we react too intensely—our children stop focusing on their own behavior and focus instead on how mean or unfair they feel we are. So do whatever you can to avoid building mountains out of molehills. Address the misbehavior and remove your child from the situation if you need to, then give yourself time to calm down before saying much, so you can be calm and thoughtful when you respond. Then you can keep the focus on your child's actions rather than your own.

9. We don't repair.

There's no way we can avoid experiencing conflict with our kids. And there's no way we'll always be on top of our game in how we handle ourselves. We'll be immature, reactive, and unkind at times. What's most important is that we address our own misbehavior and repair the breach in the relationship as soon as possible, most likely by offering and asking for forgiveness. By repairing as soon as we can in a sincere and loving manner, we model for our children a crucial skill that will allow them to enjoy much more meaningful relationships as they grow up.

10. We lay down the law in an emotional, reactive moment, then realize we've overreacted.

Sometimes our pronouncements can be a bit "supersized": "You can't go swimming for the rest of the summer!" In these moments, give yourself permission to rectify the situation. Obviously, follow-through is important or you'll lose credibility. But you can be consistent and still get out of the bind. For example, you can offer the "one more chance" card by saying, "I didn't like what you did, but I'm going to give you another try at handling things the right way." You can also admit that you overreacted: "I got mad earlier, and I wasn't thinking things through very well. I've thought about it again and I've changed my mind."

11. We forget that our children may sometimes need our help making good choices or calming themselves down.

When our kids begin to get out of control, the temptation is to demand that they "stop that right now." But sometimes, especially in the case of small children, they actually may not even be *capable* of immediately calming themselves down. That means you may need to move in and help them make good choices. The first step is to connect with your child—with both words and nonverbal communication—to help him understand that you're aware of his frustration. Only after this connection will he be prepared for you to redirect him toward making better choices. Remember, we often need to wait before responding to misbehavior. When our kids are out of control, that's not the best time to rigidly enforce a rule. When they are calmer and more receptive, they'll be better able to learn the lesson anyway.

12. We consider an audience when disciplining.

Most of us worry too much about what other people think, especially when it comes to how we parent our kids. But it's not fair to your children to discipline differently when someone else is watching. In front of in-laws, for example, the temptation might be to be harsher or more reactive because you feel that you're being judged as a parent. So remove that temptation. Pull your child aside and quietly talk to just him, without anyone else listening. Not only will this keep you from worrying how you sound to the others in the room, it will also help you get better focus from him, and you can better attune to his behavior and needs.

13. We get trapped in power struggles.

When our kids feel backed into a corner, they instinctually fight back or totally shut down. So avoid the trap. Consider giving your child an out: "Would you like to get a drink first, and then we'll pick up the toys?" Or negotiate: "Let's see if we can figure out a way for both of us to get what we need." (Obviously, there are some non-negotiables, but negotiation isn't a sign of weakness; it's a sign of respect for your child and her desires.) You can even ask your child for help: "Do you have any suggestions?" You might be shocked to find out how much your child is willing to bend in order to bring about a peaceful resolution to the standoff.

14. We discipline in response to our habits and feelings instead of responding to our individual child in a particular moment.

We sometimes lash out at our child because we're tired, or because that's what our parents did, or because we're fed up with his brother, who's been acting up all morning. It's not fair, but it's understandable. What's called for is to reflect on our behavior, to really be in the moment with our children, and to respond only to what's taking place in that instant. This is one of the most difficult tasks of parenting, but the more we can do it, the better we can respond to our kids in loving ways.

15. We embarrass our kids by correcting them in front of others.

When you have to discipline your child in public, consider her feelings. (Imagine how you'd feel if your significant other called you out on something in front of other people!) If possible, step out of the room, or just pull her close and whisper. This isn't always possible, but when you can, show your child the respect of not adding humiliation to whatever else you need to do to address the misbehavior. After all, embarrassment will just take her focus off the lesson you want to teach, and she's unlikely to hear anything you want to tell her.

16. We assume the worst before letting our kids explain.

Sometimes a situation looks bad and it really is. But sometimes things aren't as bad as they seem. Before lowering the boom, listen to your child. She may have a good explanation. It's really frustrating to believe you have a rationale for your actions, yet to have the other person say, "I don't care. I don't want to hear it. There's no reason or excuse." Obviously, you can't be naive, and any parent needs to wear her critical-thinking cap at all times. But before condemning a child for what seems obvious at first blush, find out what she has to say. Then you can decide how best to respond.

17. We dismiss our kids' experience.

When a child reacts strongly to a situation, especially when the reaction seems unwarranted and even ridiculous, the temptation is to say something like, "You're just tired," "Stop fussing," "It's not that big a deal," or "Why are you crying about this?" But statements like these minimize the child's experience. Imagine someone saying one of these phrases to you if you were upset! It's much more emotionally responsive and effective to listen, empathize, and really understand your child's experience before you respond. Even if it seems ridiculous to you, don't forget that it's very real to your child, so you don't want to dismiss something that's important to him.

18. We expect too much.

Most parents would say that they know that children aren't perfect, but most parents also expect their children to behave well all the time. Further, parents often expect too much of their children when it comes to handling emotions and making good choices—much more than is developmentally appropriate. This is especially the case with a firstborn child. The other mistake we make in expecting too much is that we assume that just because our child can handle things well sometimes, she can handle things well all the time. But especially when kids are young, their capacity to make good decisions really fluctuates. Just because they can handle things well at one time doesn't mean they can at other times.

19. We let "experts" trump our own instincts.

By "experts," we mean authors and other gurus, as well as friends and family members. It's important that we avoid disciplining our kids based on what someone else thinks we ought to do. Fill your discipline toolbox with information from lots of experts (and non-experts), then listen to your own instincts as you pick and choose different aspects of different approaches that seem to apply best to your situation with your family and your unique child.

20. We're too hard on ourselves.

We've found that it's often the most caring and conscientious parents who are too hard on themselves. They want to discipline well every time their kids mess up. But it's just not possible. So give yourself a break. Love your kids, set clear boundaries, discipline with love, and make up with them when you mess up. That kind of discipline is good for everyone involved.

The Whole-Brain Child: 12 Revolutionary Strategies to Nurture Your Child's Developing Mind

by Daniel J. Siegel, M.D., and Tina Payne Bryson, Ph.D.

You've had those days, right? When the sleep deprivation, the muddy cleats, the peanut butter on the new jacket, the homework battles, the Play-Doh in your computer keyboard, and the refrains of "She started it!" leave you counting the minutes until bedtime. On these days, when you (again?!!) have to pry a raisin from a nostril, it seems like the most you can hope for is to *survive*.

However, when it comes to your children, you're aiming a lot higher than mere survival. Of course you want to get through those difficult tantrum-in-the-restaurant moments. But whether you're a parent, grandparent, or other committed caregiver in a child's life, your ultimate goal is to raise kids in a way that lets them *thrive*. You want them to enjoy meaningful relationships, be caring and compassionate, do well in school, work hard and be responsible, and feel good about who they are.

Survive. Thrive.

We've met with thousands of parents over the years. When we ask them what matters most to them, versions of these two goals almost always top the list. They want to survive difficult parenting moments, and they want their kids and their family to thrive. As parents ourselves, we share these same goals for our own families. In our nobler, calmer, saner moments, we care about nurturing our kids' minds, increasing their sense of wonder, and helping them reach their potential in all aspects of life. But in the more frantic, stressful, bribe-the-toddler-into-the-car-seat-so-we-can-rush-to-the-soccer-game moments, sometimes all we can hope for is to avoid yelling or hearing someone say, "You're so mean!"

Take a moment and ask yourself: What do you really want for your children? What qualities do you hope they develop and take into their adult lives? Most likely, you want them to be happy, independent, and successful. You want them to enjoy fulfilling relationships and live a life full of meaning and purpose. Now think about what percentage of your time you spend intentionally developing these qualities in your children. If you're like most parents, you worry that you spend too much time just trying to get through the day (and sometimes the next five minutes) and not enough time creating experiences that help your children thrive, both today and in the future.

You might even measure yourself against some sort of perfect parent who never struggles to survive, who seemingly spends every waking second helping her children thrive. You know, the PTA president who cooks organic, fully balanced meals while reading to her kids in Latin about the importance of helping others, then escorts them to the art museum in the hybrid that plays classical music and mists lavender aromatherapy through the air-conditioning vents. None of us can match up to this imaginary superparent. Especially when we feel like a large percentage of our days is spent in full-blown survival mode, where we find ourselves wild-eyed and red-faced at the end of a birthday party, shouting, "If there's one more argument over that bow and arrow, nobody's getting *any* presents!"

If any of this sounds familiar, we've got great news for you: *the moments you are just trying to survive are actually opportunities to help your child thrive.* At times you may feel that the loving, important moments (like having a meaningful conversation about compassion or character) are separate from the parenting challenges (like fighting another homework battle or dealing with another melt-down). But they are not separate at all. When your child is disre-spectful and talks back to you, when you are asked to come in for a meeting with the principal, when you find crayon scribbles all over your wall: these are survival moments, no question about it. But at the same time, they are opportunities—even gifts—because a sur-vival moment is *also* a thrive moment, where the important, mean-ingful work of parenting takes place.

Parenting and the Brain

Parents are often experts about their children's bodies. They know that a temperature above 98.6 degrees is a fever. They know to clean out a cut so it doesn't get infected. They know which foods are most likely to leave their child wired before bedtime.

But even the best-educated, most caring parents often lack even basic information about their child's brain. Isn't this surprising? Es-pecially when you consider the central role the brain plays in virtu-ally every aspect of a child's life that parents care about: discipline, decision making, self-awareness, school, relationships, and so on. In fact, the brain pretty much determines who we are and what we do. And since the brain itself is significantly shaped by the experiences we offer as parents, knowing about the way the brain changes in re-sponse to our parenting can help us to nurture a stronger, more resil-ient brain.

So we want to introduce you to the whole-brain perspective. We'd like to explain some fundamental concepts about the brain and help you apply your new knowledge in ways that will make parenting eas-ier and more meaningful. We're not saying that the whole-brain ap-proach will get rid of all of the frustrations that come with raising kids. *But by understanding a few simple and easy-to-master basics about how the brain works, you'll be able to better understand your child, respond more effectively to difficult situations, and build a foun-dation for social, emotional, and mental health.* What you do as a parent matters, and we'll provide you with straightforward, scientifically based ideas that will help you build a strong relationship with your child that can help shape his brain well and give him the best foundation for a healthy and happy life.

What Is Integration and Why Does It Matter?

Most of us don't think about the fact that our brain has many different parts with different jobs. For example, you have a left side of the brain that helps you think logically and organize thoughts into sentences, and a right side that helps you experience emotions and read nonverbal cues. You also have a "reptile brain" that allows you to act instinctually and make split-second survival decisions, and a "mammal brain" that leads you toward connection and relationships. One part of your brain is devoted to dealing with memory; another to making moral and ethical decisions. It's almost as if your brain has multiple personalities—some rational, some irrational; some reflective, some reactive. No wonder we can seem like different people at different times!

The key to thriving is to help these parts work well together—to integrate them. Integration takes the distinct parts of your brain and helps them work together as a whole. It's similar to what happens in the body, which has different organs to perform different jobs: the lungs breathe air, the heart pumps blood, the stomach digests food. For the body to be healthy, these organs all need to be integrated. In other words, they each need to do their individual job while also working together as a whole. Integration is simply that: linking different elements together to make a well-functioning whole. Just as with the healthy functioning of the body, your brain can't perform at its best unless its different parts work together in a coordinated and balanced way. That's what integration does—it coordinates and balances the separate regions of the brain that it links together. It's easy to see when our kids aren't integrated—they become overwhelmed by their emotions, confused and chaotic. They can't respond calmly and capably to the situation at hand. Tantrums, meltdowns, aggression, and most of the other challenging experiences of parenting—and life—are a result of a loss of integration, also known as "disintegration."

We want to help our children become better integrated so they can use their whole brain in a coordinated way. For example, we want them to be *horizontally integrated,* so that their left-brain logic can work well with their right-brain emotion. We also want them to be *vertically integrated,* so that the physically higher parts of their brain, which let them thoughtfully consider their actions, work well with the lower parts, which are more concerned with instinct, gut reactions, and survival.

The way integration actually takes place is fascinating, and it's something that most people aren't aware of. In recent years, scientists have developed brain-scanning technology that allows researchers to study the brain in ways that were never before possible. This new technology has confirmed much of what we previously believed about the brain. However, one of the surprises that has shaken the very foundations of neuroscience is the discovery that the brain is actually "plastic," or moldable. This means that the brain physically changes throughout the course of our lives, not just in childhood, as we had previously assumed.

What molds our brain? Experience. Even into old age, our experiences actually change the physical structure of the brain. When we undergo an experience, our brain cells—called neurons—become active, or "fire." The brain has one hundred billion neurons, each with an average of ten thousand connections to other neurons. The ways in which particular circuits in the brain are activated determines the nature of our mental activity, ranging from perceiving sights or sounds to more abstract thought and reasoning. When neurons fire together, they grow new connections between them. Over time, the connections that result from firing lead to "rewiring" in the brain. This is incredibly exciting news. It means that we aren't held captive for the rest of our lives by the way our brain works at this moment—we can actually rewire it so that we can be healthier and happier. This is true not only for children and adolescents, but also for each of us across the life span.

Right now, your child's brain is constantly being wired and re-wired, and the experiences you provide will go a long way toward determining the structure of her brain. No pressure, right? Don't worry, though. Nature has provided that the basic architecture of the brain will develop well given proper food, sleep, and stimulation. Ge-netics, of course, play a large role in how people turn out, especially in terms of temperament. But findings from various areas in devel-opmental psychology suggest that everything that happens to us—the music we hear, the people we love, the books we read, the kind of discipline we receive, the emotions we feel—profoundly affects the way our brain develops. In other words, on top of our basic brain architecture and our inborn temperament, parents have much they can do to provide the kinds of experiences that will help develop a resilient, well-integrated brain. This book will show you how to use everyday experiences to help your child's brain become more and more integrated.

For example, children whose parents talk with them about their experiences tend to have better access to the memories of those expe-riences. Parents who speak with their children about their feelings have children who develop emotional intelligence and can under-stand their own and other people's feelings more fully. Shy children whose parents nurture a sense of courage by offering supportive ex-plorations of the world tend to lose their behavioral inhibition, while those who are excessively protected or insensitively thrust into anxiety-provoking experiences without support tend to maintain their shyness.

There is a whole field of the science of child development and at-tachment backing up this view—and the new findings in the field of neuroplasticity support the perspective that parents can directly shape the unfolding growth of their child's brain according to what experiences they offer. For example, hours of screen time—playing video games, watching television, texting—will wire the brain in certain ways. Educational activities, sports, and music will wire it in other ways. Spending time with family and friends and learning about relationships, especially with face-to-face interactions, will wire it in yet other ways. Everything that happens to us affects the way the brain develops.

This wire-and-rewire process is what integration is all about: giving our children experiences to create connections between different parts of the brain. When these parts collaborate, they create and reinforce the integrative fibers that link different parts of the brain. As a result, they are connected in more powerful ways and can work together even more harmoniously. Just as individual singers in a choir can weave their distinct voices into a harmony that would be impossible for any one person to create, an integrated brain is capable of doing much more than its individual parts could accomplish alone.

That's what we want to do for each of our kids: help their brain become more integrated so that they can use their mental resources to full capacity. With an understanding of the brain, you can be more intentional about what you teach your kids, how you respond to them, and why. You can then do much more than merely survive. By giving your children repeated experiences that develop integration, you will face fewer everyday parenting crises. But more than that, understanding integration will let you know your child more deeply, respond more effectively to difficult situations, and intentionally build a foundation for a lifetime of love and happiness. As a result, not only will your child thrive, both now and into adulthood, but you and your whole family will as well.

ACKNOWLEDGMENTS

We are profoundly grateful to all of the people who have helped shape this book that we feel so passionate about. Our teachers, colleagues, friends, students, and family members have significantly contributed to how we think about and communicate these ideas. We're especially grateful to Michael Thompson, Natalie Thompson, Janel Umfress, Darrell Walters, Roger Thompson, Gina Osher, Stephanie Hamilton, Rick Kidd, Andre van Rooyen, Lara Love, Gina Griswold, Deborah Buckwalter, Galen Buckwalter, Jay Bryson, and Liz Olson for their feedback on the book. We also thank our mentors, clinical colleagues, and the students at the Mindsight Institute and in our various seminars and parenting groups who have asked questions that push us to seek and learn more, and provided feedback about many of the ideas that make up the foundation of the No-Drama, Whole-Brain approach to parenting. There are so many people who enrich our lives and our work that we can't possibly thank you all individually, but we hope you know how much you mean to us.

We want to thank our friend and literary agent, Doug Abrams, who brought to the process not only a wealth of writing knowledge but also a passion and commitment to strengthening families and nurturing kids who are happy and healthy. We respect him as both an agent and a humanitarian. We also gratefully acknowledge the efforts and enthusiasm of our editor, Marnie Cochran, who not only offered wise counsel throughout the publication process but also extended plenty of patience as we worked to find just the right way to express the ideas so important to us. And to our fabulous illustrator, Merrilee Liddiard, we say thanks and more thanks for bringing her talent and creativity to the project and helping give the left-brain words of the book a right-brained graphic and visual life.

In addition, we thank all of the parents and patients whose stories and experiences helped us provide examples that give richness and practicality to the ideas and theories we teach. We've, of course, changed your names and the details of your stories here, but we're grateful for the power your stories lend to the communication of the No-Drama approach to discipline.

We want to acknowledge our gratitude for each other. Our collective passion for these ideas and for sharing them with the world makes working together a meaningful honor. We are grateful to our immediate and extended families who have and continue to influence who we are and celebrate what we do. Just as we have shaped who our children are and who they are becoming, they have shaped who we are as individuals and professionals, and we are deeply moved by the meaning and joy they bring to us. Finally, we thank our spouses, Caroline and Scott, who contributed in both indirect and direct ways to the production of this manuscript. They know what they mean to us, and we could never fully articulate how important to us they are, as both personal and professional partners.

Learning in life is cultivated best in our collaborative relationships with others. Our primary teachers when it comes to our own parenting have been our children—Dan's now in their twenties, Tina's in their teen and pre-teen years—who have taught us the vital importance of connection and understanding, patience and persistence. Throughout the opportunities and the challenges of being their parents, we have been reminded through their actions and re-actions, their words and their emotions, that discipline is about teaching, about learning, about finding lessons in our everyday experiences no matter how mundane or maddening. That learning is for both child and parent alike. And trying to create the necessary structure in their developing lives while parenting in a calm, even-keel, "low-drama" way has not always been easy—in fact, it is most likely one of the most challenging jobs any of us will ever have. And for these reasons, we thank both our children and our partners throughout this whole journey, for the powerful ways they each have taught us about discipline as a way of learning, of teaching, and of making life an educational adventure and a celebration of discovery. We hope this book will offer an invitation to reimagine discipline as such a learning opportunity so that you and your children will thrive and enjoy each other throughout your lives!

Dan and Tina

ABOUT THE TYPE

This book was set in Minion, a 1990 Adobe Originals typeface by Robert Slimbach (b. 1956). Minion is inspired by classical, old-style typefaces of the late Renaissance, a period of elegant, beautiful, and highly readable type designs. Created primarily for text setting, Minion combines the aesthetic and functional qualities that make text type highly readable with the versatility of digital technology.